Tensor Spaces and Exterior Algebra

Recent Titles in This Series

108 **Takeo Yokonuma,** Tensor spaces and exterior algebra, 1992

107 **B. M. Makarov, M. G. Goluzina, A. A. Lodkin, and A. N. Podkorytov,** Selected problems in real analysis, 1992

106 **G.-C. Wen,** Conformal mappings and boundary value problems, 1992

105 **D. R. Yafaev,** Mathematical scattering theory: General theory, 1992

104 **R. L. Dobrushin, R. Kotecký, and S. Shlosman,** Wulff construction: A global shape from local interaction, 1992

103 **A. K. Tsikh,** Multidimensional residues and their applications, 1992

102 **A. M. Il'in,** Matching of asymptotic expansions of solutions of boundary value problems, 1992

101 **Zhang Zhi-fen, Ding Tong-ren, Huang Wen-zao, and Dong Zhen-xi,** Qualitative theory of differential equations, 1992

100 **V. L. Popov,** Groups, generators, syzygies, and orbits in invariant theory, 1992

99 **Norio Shimakura,** Partial differential operators of elliptic type, 1992

98 **V. A. Vassiliev,** Complements of discriminants of smooth maps: Topology and applications, 1992

97 **Itiro Tamura,** Topology of foliations: An introduction, 1992

96 **A. I. Markushevich,** Introduction to the classical theory of Abelian functions, 1992

95 **Guangchang Dong,** Nonlinear partial differential equations of second order, 1991

94 **Yu. S. Il'yashenko,** Finiteness theorems for limit cycles, 1991

93 **A. T. Fomenko and A. A. Tuzhilin,** Elements of the geometry and topology of minimal surfaces in three-dimensional space, 1991

92 **E. M. Nikishin and V. N. Sorokin,** Rational approximations and orthogonality, 1991

91 **Mamoru Mimura and Hirosi Toda,** Topology of Lie groups, I and II, 1991

90 **S. L. Sobolev,** Some applications of functional analysis in mathematical physics, third edition, 1991

89 **Valeriĭ V. Kozlov and Dmitriĭ V. Treshchëv,** Billiards: A genetic introduction to the dynamics of systems with impacts, 1991

88 **A. G. Khovanskiĭ,** Fewnomials, 1991

87 **Aleksandr Robertovich Kemer,** Ideals of identities of associative algebras, 1991

86 **V. M. Kadets and M. I. Kadets,** Rearrangements of series in Banach spaces, 1991

85 **Mikio Ise and Masaru Takeuchi,** Lie groups I, II, 1991

84 **Đáo Trọng Thi and A. T. Fomenko,** Minimal surfaces, stratified multivarifolds, and the Plateau problem, 1991

83 **N. I. Portenko,** Generalized diffusion processes, 1990

82 **Yasutaka Sibuya,** Linear differential equations in the complex domain: Problems of analytic continuation, 1990

81 **I. M. Gelfand and S. G. Gindikin, Editors,** Mathematical problems of tomography, 1990

80 **Junjiro Noguchi and Takushiro Ochiai,** Geometric function theory in several complex variables, 1990

79 **N. I. Akhiezer,** Elements of the theory of elliptic functions, 1990

78 **A. V. Skorokhod,** Asymptotic methods of the theory of stochastic differential equations, 1989

77 **V. M. Filippov,** Variational principles for nonpotential operators, 1989

76 **Phillip A. Griffiths,** Introduction to algebraic curves, 1989

75 **B. S. Kashin and A. A. Saakyan,** Orthogonal series, 1989

74 **V. I. Yudovich,** The linearization method in hydrodynamical stability theory, 1989

(Continued in the back of this publication)

Translations of
MATHEMATICAL MONOGRAPHS

Volume 108

Tensor Spaces and Exterior Algebra

Takeo Yokonuma

AMERICAN MATHEMATICAL SOCIETY
Providence, Rhode Island

テンソル空間と外積代数

TENSORU KUUKAN TO GAISEKIDAISUU
(Tensor Spaces and Exterior Algebra)
by Takeo Yokonuma

Originally published in Japanese by Iwanami Shoten, Publishers, Tokyo in 1977
Translated from the Japanese by Takeo Yokonuma

2000 *Mathematics Subject Classification.* Primary 15A69, 47-XX;
Secondary 15A75, 81-XX.

ABSTRACT. This book provides an introduction to tensors and related topics. The book begins with definitions of the basic concepts of the theory; tensor products of vector spaces, tensors, tensor algebras, and exterior algebra. Their properties are then studied and applications given.

Algebraic systems with bilinear multiplication are introduced in the final chapter. In particular, the theory of replicas of Chevalley and several properties of Lie algebras that follow from this theory are presented.

Library of Congress Cataloging-in-Publication Data

Yokonuma, Takeo, 1939–
[Tensoru kuukan to gaisekidaisuu. English]
Tensor spaces and exterior algebra/Takeo Yokonuma; translated by Takeo Yokonuma.
p. cm.—(Translations of mathematical monographs, ISSN 0065-9282; 108)
Translation of: Tensoru kuukan to gaisekidaisuu.
Includes bibliographical references and index.
ISBN 0-8218-4564-0
1. Multilinear algebra. 2. Tensor products. I. Title. II. Series.
QA199.5.Y6513 1992 92-16721
512′.57–dc20 CIP

AMS softcover ISBN 978-0-8218-2796-3

The paper used in this book is acid-free and falls within the guidelines
established to ensure permanence and durability. ♾
This publication was typeset using $\mathcal{A}\!\mathcal{M}\!\mathcal{S}$-TEX,
the American Mathematical Society's TEX macro system.
Visit the AMS home page at URL: `http://www.ams.org/`

10 9 8 7 6 5 4 3 2 06 05 04 03 02 01

Contents

Preface to the English Edition

This is a translation of my work originally published by Iwanami Shoten, as a volume in their Lecture Series on linear algebra.

We assume, therefore, that readers are familiar with several fundamental concepts of linear algebra such as vector spaces, matrices, determinants, etc., though we review some of these concepts in the text. We have made some changes in the references.

On this occasion, I would like to express my gratitude to Professor Nagayoshi Iwahori for giving me encouragement and many valuable suggestions during the preparation of the original work and also to Mr. Hideo Arai of Iwanami Shoten for continued support.

I am grateful to the American Mathematical Society and the staff for their effort in publishing this English edition. I also thank Kazunari Noda and Nami Yokonuma for their excellent typing.

Takeo Yokonuma
December 1991

Preface

The subject matter of the present book is generally called multilinear algebra; we shall discuss mainly tensors and related concepts. The readers may recall several important tensors that are used in differential geometry, mechanics, electromagnetics, and so on. On the other hand, it is generally said that tensors are difficult to understand. One of the reasons for this is that formerly tensors and tensor fields (mappings whose values are tensors) were not distinguished, and tensor fields were discussed without defining tensors in advance.([1]) In fact, readers should be aware that sometimes tensor fields are simply called tensors in the literature. In any case, it is important to understand clearly what tensors are. Our purpose is to explain tensors to the readers as clearly as possible. Tensor fields are defined by means of tensors. The analytic theory of tensor fields is known as tensor calculus. Needless to say, it is beyond the scope of linear algebra and we shall not discuss it here. The notions of tensor density and pseudotensor are also important in applications of tensors. Since they are related to representations of linear groups, we shall mention only their definitions (§II.6).

As stated above, the notion of tensor seems to have originated from what are now called tensor fields. Research has been done, since the end of last century, with the study of vector calculus and the theory of invariants, on systems of functions which satisfy certain transformation laws. Ricci and Levi-Civita founded tensor calculus. It became well known after being used by Einstein to describe the theory of relativity. When we try to describe a physical phenomenon, we use a coordinate system. Though the description depends on the coordinate system, the phenomenon itself does not. To explain the situation, systems of functions that satisfy a certain transformation law play an important role. Depending on the type of transformation law, several kinds of tensor (field) have been defined. By the way, it seems that the word "tensor" was derived from tension. Thus, a tensor is classically defined as a system (or an array) of numbers which satisfies a certain transformation law (see §II.2). We can consider that this system describes an "object" whose existence does not depend on coordinate systems. This point of view

([1])See R. Godement: Cours d'algèbre, Hermann, p. 269.

is sufficient for computation but it does not explain what the object is.

In order to define tensors, we begin in Chapter I by constructing tensor products of vector spaces. Then, in Chapter II, we define tensors and study their properties. In many branches of mathematics the construction of tensor products is used as a powerful tool to construct new objects from known ones. For example, we will describe the extension of a field of scalars (§I.8b). The tensor product of representations is another example.

In Chapter III, we discuss the notion of exterior algebra. Exterior algebras are basic to the theory of differential forms. In Chapter IV, to show another aspect of the theory of tensor products, we discuss algebraic systems with bilinear multiplication. In particular, we discuss Lie algebras.

CHAPTER I

Definition of Tensor Products

In this chapter, we explain in detail the notion of tensor product of vector spaces. Using this notion, we can explicitly define tensors, as will be done in Chapter II.

The tensor product of vector spaces is a new vector space which is associated with given vector spaces in a certain natural way. (To simplify the presentation, we describe the case of two vector spaces, say V and W. In general, we can define the tensor product of any finite number of vector spaces.) Recall the direct sum of vector spaces. This is another example of a new vector space associated with given vector spaces in a natural way. In the case of the direct sum, we can easily construct the direct sum $V \oplus W$, starting from vector spaces V and W. However, in the case of the tensor product, the construction is not so easy to understand. So we make a detour in a sense. First, we prove (§3, Theorem 1.1) the existence and the uniqueness of a vector space which has the following property—every bilinear mapping on $V \times W$ is "linearized on this space." Then, we call this vector space a tensor product of V and W. The proof of existence is given in such a way that we can manage the tensor product. By uniqueness, we can define the tensor product without ambiguity.

To formulate the results, bilinear mapping is defined in §2 and the tensor product is defined in §4. The various properties of the tensor product are then explained in subsequent sections.

§1. Preliminaries

First we recall notions that will be used frequently in this book. Throughout this book, we assume that every field has characteristic 0. By definition a field k has characteristic 0 if the sum $n \cdot 1 = 1 + \cdots + 1$ (n terms) is not equal to 0 for any natural number n, where 1 is the unit element of k. The field of rational numbers $\mathbf{Q}$, the field of real numbers $\mathbf{R}$, and the field of complex numbers $\mathbf{C}$ have characteristic 0. On the other hand, the finite field with p elements does not have characteristic 0, since $p \cdot 1 = 1 + \cdots + 1$ is equal to 0.

We also assume that all vector spaces over a field k are finite dimensional unless otherwise stated. We denote the dimension of V by $\dim V$ or $\dim_k V$. A vector space over a field k is sometimes called a k-vector space.

Let V be a vector space over a field k. The *dual space* V^* of V is the k-vector space consisting of all linear mappings of V into k. The sum $\varphi + \psi$ and the scalar multiple $\alpha\varphi$ for $\varphi, \psi \in V^*$, $\alpha \in k$ are defined by the following formulas.

$$(\varphi + \psi)(v) = \varphi(v) + \psi(v), \quad (\alpha\varphi)(v) = \alpha(\varphi(v)) \qquad (v \in V).$$

Let $\mathscr{E} = (e_1, \ldots, e_n)$ $(n = \dim V)$ be a basis for V. For $i = 1, \ldots, n$, consider the element $e_i^* \in V^*$ defined by $e_i^*(e_j) = \delta_{ij}$ (1). Then the elements $e_1^*, \ldots, e_n^*$ form a basis for V^*, that is called the *dual basis* of $\mathscr{E}$. In particular, we have $\dim V^* = \dim V$.

Let v be an element of V. Then the mapping which assigns $\varphi(v)$ to $\varphi \in V^*$ is a linear mapping of V^* into k, which we denote by v^0. By $v \mapsto v^0$, to every element v of V there corresponds an element of $(V^*)^*$. We can verify that this correspondence is linear and bijective. Therefore (2) $V \cong (V^*)^*$ by this canonical mapping $v \mapsto v^0$.

Let V and W be k-vector spaces. The set of linear mappings of V into W is also a vector space over k. This set is denoted by $\mathrm{Hom}(V, W)$. For $F, G \in \mathrm{Hom}(V, W)$, $\alpha \in k$, the sum and scalar multiplication are defined by the following formulas.

$$(F + G)(v) = F(v) + G(v), \quad (\alpha F)(v) = \alpha(F(v)) \qquad (v \in V).$$

To emphasize the field k, we sometimes use the term k-linear mapping and the notation $\mathrm{Hom}_k(V, W)$.

Let $(e_1, \ldots, e_n)$ be a basis for V and $(f_1, \ldots, f_m)$ a basis for W $(n = \dim V,\ m = \dim W)$. A linear mapping $F \in \mathrm{Hom}(V, W)$ is determined by n elements $F(e_j)$ $(1 \le j \le n)$ of W. And, for n arbitrary elements w_j $(1 \le j \le n)$ of W, there exists one and only one linear mapping $G \in \mathrm{Hom}(V, W)$ such that $G(e_j) = w_j$ for all j. For all i and j $(1 \le i \le m,\ 1 \le j \le n)$ consider the elements F_{ij} defined by

$$F_{ij}(e_j) = f_i, \quad F_{ij}(e_k) = 0 \qquad (1 \le k \le n,\ k \ne j).$$

Then F_{ij} $(1 \le i \le n,\ 1 \le j \le m)$ form a basis for $\mathrm{Hom}(V, W)$. In particular, $\dim \mathrm{Hom}(V, W) = nm = \dim V \cdot \dim W$.

§2. Definition of bilinear mappings

Among the mappings between vector spaces the linear ones are important. Bilinear mappings are defined by extending the definition of linear mappings. First we recall the definition of a linear mapping. Let V and W be k-vector

(1) δ_{ij}, called Kronecker's symbol, is defined by $\delta_{ij} = 1$ if $i = j$ and $\delta_{ij} = 0$ if $i \ne j$.

(2) $A \cong B$ denotes that A and B are isomorphic.

spaces. A mapping $F: V \to U$ that assigns $F(v) \in U$ to $v \in V$ is called a *linear mapping* if it has the following property (L):

(L) For all $\alpha, \beta \in k$, $v_1, v_2 \in V$, $F(\alpha v_1 + \beta v_2) = \alpha F(v_1) + \beta F(v_2)$.

A bilinear mapping is an extension of the notion of linear mapping of two variables. Let V, W, and U be k-vector spaces. Consider a mapping $\Phi: V \times W \to U$ that, to $v \in V$ and $w \in W$, assigns $\Phi(v, w) \in U$.

DEFINITION 1.1. A mapping $\Phi: V \times W \to U$ is called a *bilinear mapping* if it is linear in each variable. Namely, Φ has the following properties (1) and (2).

(1) For any fixed element w_0 of W, the mapping of V into U defined by $v \mapsto \Phi(v, w_0)$ is a linear mapping.

(2) For any fixed element v_0 of V, the mapping of W into U defined by $w \mapsto \Phi(v_0, w)$ is a linear mapping.

By virtue of (L), (1) and (2) are restated as follows:

For all $\alpha, \beta \in k$, $v_0, v_1, v_2 \in V$, $w_0, w_1, w_2 \in W$,

$$\Phi(\alpha v_1 + \beta v_2, w_0) = \alpha\Phi(v_1, w_0) + \beta\Phi(v_2, w_0),$$
$$\Phi(v_0, \alpha w_1 + \beta w_2) = \alpha\Phi(v_0, w_1) + \beta\Phi(v_0, w_2).$$

If U is the 1-dimensional space k, a bilinear mapping $V \times W \to k$ is called a *bilinear form*.

EXAMPLE 1.1. We denote by $M(m, n; k)$ the set of $m \times n$ matrices with coefficients in k. $M(m, n; k)$ is a vector space over k with respect to the usual addition and scalar multiplication. Let $V = M(m, l; k)$, $W = M(l, n; k)$, and $U = M(m, n; k)$. Define $\Phi(A, B) = AB$ (the product of matrices A and B) for $A \in V$, $B \in W$. Then Φ is a bilinear mapping of $V \times W$ into U.

EXAMPLE 1.2. Let V, W be vector spaces and let V^*, W^* be the dual spaces of V and W respectively. For fixed $\varphi \in V^*$ and $\psi \in W^*$, the mapping $V \times W \ni (v, w) \mapsto \varphi(v)\psi(w) \in k$ is a bilinear form on $V \times W$. Similarly, for fixed $(v, w) \in V \times W$, the mapping $V^* \times W^* \ni (\varphi, \psi) \mapsto \varphi(v)\psi(w)$ is a bilinear form on $V^* \times W^*$.

We denote by $\mathscr{L}(V, W; U)$ the set of bilinear mappings of $V \times W$ into U. Then, as in the case of linear mappings we can easily show the following properties:

(1) For $\Phi_1, \Phi_2 \in \mathscr{L}(V, W; U)$, $\alpha, \beta \in k$, define

$$(\alpha\Phi_1 + \beta\Phi_2)(v, w) = \alpha\Phi_1(v, w) + \beta\Phi_2(v, w) \qquad (v \in V,\ w \in W).$$

Then $\alpha\Phi_1 + \beta\Phi_2 \in \mathscr{L}(V, W; U)$. With respect to this operation $\mathscr{L}(V, W; U)$ is a vector space over k.

(2) Let $(e_1, \dots, e_n)$ and $(f_1, \dots, f_m)$ be bases for V and W respectively. Then, an element Φ of $\mathscr{L}(V, W; U)$ is determined by mn elements $\Phi(e_i, f_j)$ $(1 \le i \le n,\ 1 \le j \le m)$ of U. And, for arbitrary nm elements u_{ij} of U, there exists $\Psi \in \mathscr{L}(V, W; U)$ such that $\Psi(e_i, f_j) = u_{ij}$.

In fact, for $v = \sum_i \alpha_i e_i$, $w = \sum_j \beta_j f_j$, we have

$$\Phi(v, w) = \sum_{i,j} \alpha_i \beta_j \Phi(e_i, f_j).$$

On the other hand, for given u_{ij}, define a mapping $\Psi: V \times W \to U$ by $\Psi(v, w) = \sum_{i,j} \gamma_i \delta_j u_{ij}$, where $v = \sum_i \gamma_i e_i \in V$, $w = \sum_j \delta_j f_j \in W$. Clearly, Ψ is bilinear and $\Psi(e_i, f_j) = u_{ij}$.

(3) Moreover, let $(g_1, \dots, g_l)$ be a basis for U. For all indices a, b, c $(1 \le a \le n,\ 1 \le b \le m,\ 1 \le c \le l)$, define $\Phi_{abc} \in \mathscr{L}(V, W; U)$ by

$$\Phi_{abc}(e_i, f_j) = \begin{cases} g_c & \text{when } i = a \text{ and } j = b, \\ 0 & \text{otherwise.} \end{cases}$$

These Φ_{abc} then form a basis for the k-vector space $\mathscr{L}(V, W; U)$. In particular, $\dim \mathscr{L}(V, W; U) = \dim V \cdot \dim W \cdot \dim U$.

EXERCISE 1. By condition (2) of Definition 1.1, for a fixed element of V, elements of $\mathscr{L}(V, W; U)$ define linear mappings of W into U. Using this remark, prove that

$$\mathscr{L}(V, W; U) \cong \mathrm{Hom}(V, \mathrm{Hom}(W, U)).$$

Similarly, prove that

$$\mathscr{L}(V, W; U) \cong \mathrm{Hom}(W, \mathrm{Hom}(V, U)).$$

EXERCISE 2. Let Φ be an element of $\mathscr{L}(V, W; U)$. The set of images of Φ,

$$\mathrm{Im}\,\Phi = \Phi(V \times W) = \{\Phi(v, w) \mid v \in V,\ w \in W\}$$

is not necessarily a (vector) subspace. Give an example of Φ such that $\mathrm{Im}\,\Phi$ is not a (vector) subspace. (Compare with the case of linear mappings.)

§3. Linearization of bilinear mappings

Let V and W be k-vector spaces. Consider the sets $\mathscr{L}(V, W; U)$ of bilinear mappings of $V \times W$ into U for various k-vector spaces U. It is known that, among U, there exist a k-vector space U_0 and a bilinear mapping $\iota: V \times W \to U_0$ which have a particular property. Here, "particular" means that, through U_0 and ι, we can "linearize" all $\Phi \in \mathscr{L}(V, W; U)$. More precisely, we have the following theorem.

THEOREM 1.1. (1) *Let V and W be k-vector spaces. There exist a k-vector space U_0 and a bilinear mapping $\iota \in \mathscr{L}(V, W; U_0)$, which satisfy the following conditions* (T1) *and* (T2).

(T1) *U_0 is generated by the image $\iota(V \times W)$ of ι.*

(T2) *For any $\Phi \in \mathscr{L}(V, W; U)$, there exists a k-linear mapping $F: U_0 \to U$ such that $\Phi = F \circ \iota$.*

$$\begin{array}{ccc} V \times W & \xrightarrow{\ \iota\ } & U_0 \\ & \searrow^{\Phi} & \downarrow F \\ & & U \end{array}$$

(2) *The pair* (U_0, ι) *is unique in the following sense: If the pairs* (U_0, ι) *and* (U_0', ι') *consisting of a* k*-vector space and a* k*-bilinear mapping satisfy conditions* (T1) *and* (T2), *then there exists a unique linear isomorphism* $F_0 \colon U_0 \to U_0'$ *such that* $F_0 \circ \iota = \iota'$.

NOTE. (For the reader's convenience, elementary but important facts are recalled in the notes.) Let X be a vector space and S be a subset of X. The intersection of all subspaces of X containing S is also a subspace containing S and is the smallest among them. This space is called the subspace generated (or spanned) by S and is denoted by $\langle S \rangle$. It is easy to see that $\langle S \rangle$ is the set of all finite linear combination of elements of S. When $\langle S \rangle = X$, X is said to be generated by S and S is called a set of generators (or a generating set) of X.

Before giving the proof of the theorem, we restate the condition on (U_0, ι). The condition (T1) is equivalent to the uniqueness of F in (T2). Namely, (T1) and (T2) are equivalent to the following (T).

(T) For any $\Phi \in \mathscr{L}(V, W; U)$, there exists one and only one k-linear mapping $F \colon U_0 \to U$ such that $\Phi = F \circ \iota$.

Suppose that (U_0, ι) satisfies (T1) and (T2). The existence of F follows from (T2). Suppose that F and F' are linear mappings $U_0 \to U$ such that $\Phi = F \circ \iota = F' \circ \iota$. Since F and F' are linear mappings that coincide on the generating set $\iota(V \times W)$ of U_0, we have $F = F'$, which shows that (U_0, ι) satisfies (T).

Conversely, suppose that (U_0, ι) satisfies (T). Clearly we have (T2). Let U_0' be the subspace of U_0 generated by $\iota(V \times W)$. Since the image of ι is contained in U_0', ι can be considered as a mapping of $V \times W$ into U_0', which we denote by ι_1. Applying (T2) to $\Phi = \iota_1$, we have a linear mapping $F \colon U_0 \to U_0'$ such that $\iota_1 = F \circ \iota$. Let j be the inclusion mapping of U_0' into U. Then $\iota = j \circ \iota_1$, therefore we have $\iota = j \circ \iota_1 = j \circ F \circ \iota$. On the other hand, the identity mapping id of U_0 ($\mathrm{id}(u) = u$ for $u \in U_0$) clearly satisfies $\mathrm{id} \circ \iota = \iota$, which shows that $F = \mathrm{id}$ satisfies (T) for $\Phi = \iota$. Since $(j \circ F) \circ \iota = \iota$, we have $j \circ F = \mathrm{id}$ by the uniqueness of F in (T). In particular, it follows that j is surjective and $U_0' = U$, which implies (T1).

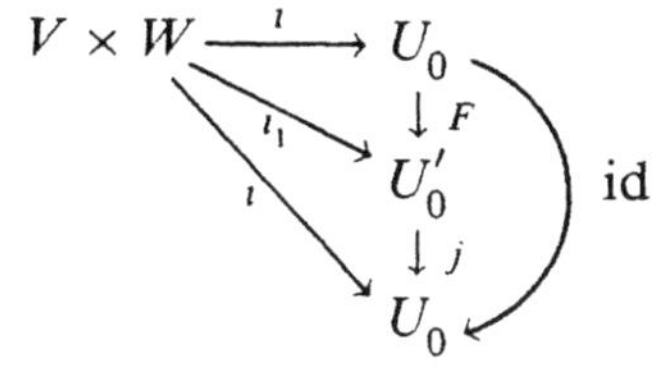

PROOF OF UNIQUENESS IN THEOREM 1.1. We first prove (2), which is easier than (1). Assume that (U_0, ι) and (U_0', ι') have the property mentioned in (1). Since ι' is a bilinear mapping: $V \times W \to U_0'$, applying (T) to (U_0, ι) we have a linear mapping $F_0 \colon U_0 \to U_0'$ such that $F_0 \circ \iota = \iota'$. Similarly, we have a linear mapping $G_0 \colon U_0' \to U_0$ such that $G_0 \circ \iota' = \iota$.

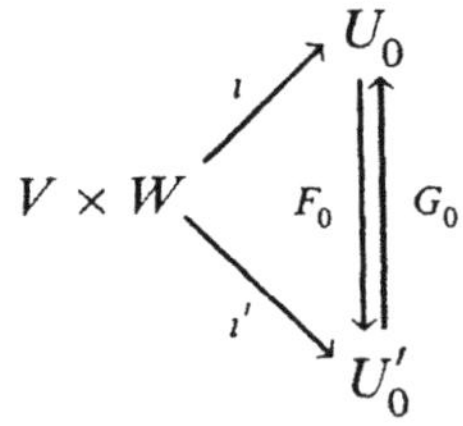

Then $\iota = G_0 \circ \iota' = (G_0 \circ F_0) \circ \iota$. For the identity mapping id of U_0, clearly we have $\mathrm{id} \circ \iota = \iota$. From the uniqueness, $G_0 \circ F_0 = \mathrm{id}$. Similarly, for the identity mapping id$'$ of U_0', we have $F_0 \circ G_0 = \mathrm{id}'$, which shows that F_0 is an isomorphism. The uniqueness of F_0 follows from uniqueness in (T). □

Before proving existence, we are going to analyze what conditions are necessary for U_0.

The relation $\Phi = F \circ \iota$ implies that the image $\Phi(V \times W)$ of Φ is contained in $F(U_0)$, which shows that the subspace $\langle \Phi(V \times W) \rangle$ generated by $\Phi(V \times W)$ is contained in $F(U_0)$. Set $\dim V = n$, $\dim W = m$ and take bases $\mathscr{E} = (e_1, \dots, e_n)$ and $\mathscr{F} = (f_1, \dots, f_m)$ for V and W respectively. By property (2) of a bilinear mapping, we can choose nm elements $\Phi(e_i, f_j)$ arbitrarily. Therefore we can choose Φ and U in such a way that $\{\Phi(e_i, f_j) \mid 1 \le i \le n,\ 1 \le j \le m\}$ is linearly independent. In this case, we have $\dim \langle \Phi(U \times V) \rangle = nm$ and $\dim F(U_0) \ge nm$. Since $\dim U_0 \ge \dim F(U_0)$, we must have $\dim U_0 \ge nm$. In fact, we shall prove that any vector space of $\dim nm$ satisfies the conditions (T1) and (T2) for an appropriate ι.

PROOF OF EXISTENCE IN THEOREM 1.1. Let U_0 be a vector space of $\dim mn$. Take a basis $\mathscr{G} = (g_1, \dots, g_{mn})$ of U_0. Arrange the elements of $\mathscr{G}$ in n rows where each row has m elements and set $g_l = g(i, j)$ if g_l is in ith row and jth position in that row $(1 \le l \le mn,\ 1 \le i \le n,\ 1 \le j \le m)$. Thus we have a one-to-one correspondence $(e_i, f_j) \leftrightarrow g(i, j)$ between the sets $\mathscr{E} \times \mathscr{F}$ and $\mathscr{G}$. Define a mapping $\iota \colon V \times W \to U_0$ by

$$\iota(v, w) = \sum_{i,j} \alpha_i \beta_j g(i, j),$$

where $v = \sum_i \alpha_i e_i$, $w = \sum_j \beta_j f_j$ $(\alpha_i, \beta_j \in k)$.

We will show that the above (U_0, ι) satisfies (T1) and (T2). Clearly ι is a bilinear mapping and (T1) follows from $\iota(e_i, f_j) = g(i, j)$. To prove (T2), let Φ be an element of $\mathscr{L}(V, W; U)$ for a vector space U. Define $F_\Phi \colon U_0 \to U$ by

$$F_\Phi(u) = F_\Phi\left(\sum_{i,j} \gamma_{ij} g(i, j)\right) = \sum_{i,j} \gamma_{ij} \Phi(e_i, f_j),$$

where $u = \sum_l \gamma_l g_l = \sum_{i,j} \gamma_{ij} g(i, j)$ $(\gamma_{ij} = \gamma_l \in k$ if $g(i, j) = g_l)$. Then the mapping F_Φ is linear and $\Phi = F_\Phi \circ \iota$. □

REMARK. Property (T) of (U_0, ι) can be interpreted as stating that (U_0, ι) dominates all the elements of $\mathscr{L}(V, W; U)$. Such a property is called "universality." Later we will give several additional examples of universality.

§4. Definition of tensor products

We are now ready to define the tensor product of k-vector spaces.

DEFINITION 1.2. Let V and W be k-vector spaces. The pair (U_0, ι) consisting of a k-vector space U_0 and a bilinear mapping $\iota: V \times W \to U_0$, satisfying (T1) and (T2), the existence of which is assured by Theorem 1.1, is called a *tensor product* of V and W. We write $U_0 = V \otimes W$ and $\iota(v, w) = v \otimes w$. The mapping ι is called the *canonical mapping* of a tensor product $U_0 = V \otimes W$.

In the following, we sometimes say that a vector space U_0 is a tensor product $V \otimes W$ of V and W. Implicitly this means that there exists a bilinear mapping $\iota: V \times W \to U_0$ satisfying (T1) and (T2) (or (T)). In this notation, condition (T) can be restated as follows: a tensor product U_0 of V and W is generated by $\{v \otimes w\ (= \iota(v, w)) \mid v \in V,\ w \in W\}$. The uniqueness property (2) of Theorem 1.1 can be restated as follows: if U_0 and U_0' are tensor products of V and W, then there exists a unique linear isomorphism $F_0: U_0 \to U_0'$ such that F_0 associates $v \otimes w$ in U_0' to $v \otimes w$ $(= \iota(v, w))$ in U_0 for all $v \in V$, $w \in W$.

In the proof of existence in Theorem 1.1, we used bases for V and W. Therefore, it might be difficult to understand the meaning of the tensor product. Thus, we give another construction of a tensor product (U_0, ι) free from bases.

Let V^* and W^* be the dual spaces of V and W respectively (see §1 for dual spaces), and let U_0 be defined by

$$U_0 = \mathscr{L}(V^*, W^*; k).$$

For fixed $v \in V$ and $w \in W$, the mapping $V^* \times W^* \to k$ which assigns $\varphi(v)\psi(w)$ to $(\varphi, \psi) \in V^* \times W^*$ is clearly bilinear (cf. §2, Example 1.2). So it is an element of U_0, which we denote by $\iota(v, w)$. Thus we have a mapping $\iota: V \times W \to U_0$, which is also bilinear. (Check this.) Then we can show that (U_0, ι) satisfies conditions (T1) and (T2) of Theorem 1.1. Take bases $(e_1, \dots, e_n)$ for V and $(f_1, \dots, f_m)$ for W. Let $(e_1^*, \dots, e_n^*)$ and $(f_1^*, \dots, f_m^*)$ be their dual bases. We can construct a basis (Φ_{ij}) for U_0, where Φ_{ij} $(1 \le i \le n,\ 1 \le j \le m)$ are defined as follows:

$$\Phi_{ij}(e_a^*, f_b^*) = \begin{cases} 1 & \text{if } a = i,\ b = j, \\ 0 & \text{otherwise.} \end{cases}$$

Note that

$$\Phi_{ij} = \iota(e_i, f_j),$$

from which we obtain condition (T1). For every $\Phi \in \mathscr{L}(V, W; U)$, define a linear mapping $F_\Phi: U_0 \to U$ by

$$F_\Phi\left(\sum_{i,j} \gamma_{ij}\Phi_{ij}\right) = \sum_{i,j} \gamma_{ij}\Phi(e_i, f_j) \qquad (\gamma_{ij} \in k).$$

Then $\Phi = F_\Phi \circ \iota$, which implies condition (T2).

Thus we can say that the tensor product $V \otimes W$ is simply $\mathscr{L}(V^*, W^*; k)$ and that, in this situation, $v \otimes w$ is equal to the bilinear mapping $\iota(v, w)$.

§5. Properties of tensor products

We begin with properties which can be easily obtained. Rewriting the bilinearity of the canonical mapping ι, we have the following proposition.

PROPOSITION 1.1 (bilinearity of $\otimes$). *For $\alpha, \beta \in k$, $v, v_1, v_2 \in V$, $w, w_1, w_2 \in W$, we have*

$$(\alpha v_1 + \beta v_2) \otimes w = \alpha(v_1 \otimes w) + \beta(v_2 \otimes w),$$
$$v \otimes (\alpha w_1 + \beta w_2) = \alpha(v \otimes w_1) + \beta(v \otimes w_2).$$

From the proof of existence of tensor products (§3) we obtain

PROPOSITION 1.2. *Let $(e_1, \dots, e_n)$ be a basis for V and $(f_1, \dots, f_m)$ a basis for W. Then the mn elements $e_i \otimes f_j$ $(1 \le i \le n,\ 1 \le j \le m)$ form a basis for $V \otimes W$.*

In particular,

$$\dim(V \otimes W) = \dim V \cdot \dim W.$$

COROLLARY. *Let $v \in V$ and $w \in W$ be nonzero elements. Then $v \otimes w \neq 0$.*

PROOF. Take bases of V and W which contain v and w respectively. Then $v \otimes w$ is a basis element of $V \otimes W$. Therefore $v \otimes w \neq 0$. □

If we arrange the basis elements $e_i \otimes f_j$ in the order $e_1 \otimes f_1, \dots, e_1 \otimes f_m$, $e_2 \otimes f_1, \dots, e_2 \otimes f_m, \dots, e_n \otimes f_1, \dots, e_n \otimes f_m$, then we say that the basis elements are arranged in *lexicographical order*.

Since $V \otimes W$ is generated by $\iota(V \times W)$, every element of $V \otimes W$ can be expressed as a sum $\sum_{i=1}^r (v_i \otimes w_i)$ of elements $v_i \otimes w_i = \iota(v_i, w_i)$, $v_i \in V$, $w_i \in W$. In general, an element has many expressions of this form and we may choose convenient ones among them according to the situation. Since $\iota(V \times W) \neq V \otimes W$ in general, not every element can be written in the form $v \otimes w$. Let us give some examples. Proposition 1.2 implies that every element can be expressed uniquely in the form $\sum_{i,j} \alpha_{ij}(e_i \otimes f_j)$ $(\alpha_{ij} \in k)$. Then we have

$$\sum_{i,j} \alpha_{ij}(e_i \otimes f_j) = \sum_i \left\{ e_i \otimes \left(\sum_j \alpha_{ij} f_j \right) \right\} = \sum_j \left\{ \left(\sum_i \alpha_{ij} e_i \right) \otimes f_j \right\}.$$

LEMMA 1.1. *Let* $u = \sum_{i=1}^{r}(v_i \otimes w_i) \in V \otimes W$, *where* $v_i \in V$, $w_i \in W$. *If* $\{v_1, \dots, v_r\}$ *is linearly independent, then the elements* $w_1, \dots, w_r$ *are uniquely determined. Namely, if* $u = \sum_{i=1}^{r}(v_i \otimes w_i) = \sum_{i=1}^{r}(v_i \otimes w_i')$, *we have* $w_i = w_i'$ *for all* $i = 1, 2, \dots, r$. *Similarly, if* $\{w_1, \dots, w_r\}$ *is linearly independent, the elements* $v_1, \dots, v_r$ *are uniquely determined.*

PROOF. Let us prove the first assertion. Choose a basis $(v_1, \dots, v_r, v_{r+1}, \dots, v_n)$ for V that contains the linearly independent elements $v_1, \dots, v_r$. Since $\sum_{i=1}^{r}(v_i \otimes w_i) = \sum_{i=1}^{r}(v_i \otimes w_i')$, we have $\sum_{i=1}^{r}\{v_i \otimes (w_i - w_i')\} = 0$ by Proposition 1.1. Take any basis $(f_1, \dots, f_m)$ for W. Set $w_i - w_i' = \sum_{j=1}^{m}\beta_{ij}f_j$. Then

$$\sum_{i=1}^{r}\{v_i \otimes (w_i - w_i')\} = \sum_{i=1}^{r}\left\{v_i \otimes \left(\sum_{j=1}^{m}\beta_{ij}f_j\right)\right\} = \sum_{i=1}^{r}\sum_{j=1}^{m}\beta_{ij}(v_i \otimes f_j) = 0.$$

Since $(v_i \otimes f_j)$ is a basis for $V \otimes W$, this implies $\beta_{ij} = 0$ for all i and j, i.e., $w_i = w_i'$. □

By property (T) of the tensor product, for any $\Phi \in \mathscr{L}(V, W; U)$ there exists a unique $F \in \mathrm{Hom}(V \otimes W, U)$ such that $\Phi = F \circ \iota$. This correspondence gives the next proposition.

PROPOSITION 1.3. *As k-vector spaces,*

$$\mathscr{L}(V, W; U) \cong \mathrm{Hom}(V \otimes W, U).$$

PROOF. For the proof of this assertion, we give the inverse of the correspondence above and show that it is a linear isomorphism. For $F \in \mathrm{Hom}(V \otimes W, U)$, define $\Phi_F = F \circ \iota$. Then $\Phi_F \in \mathscr{L}(V, W; U)$. Property (T) implies that the mapping $F \mapsto \Phi_F$ is bijective. The fact that this map is linear, namely,

$$\Phi_{\alpha F} = \alpha\Phi_F, \qquad \Phi_{F+F'} = \Phi_F + \Phi_{F'}$$

for $F, F' \in \mathrm{Hom}(V \otimes W, U)$, $\alpha \in k$, follows easily from the definition of Φ. □

Recall that the dual space of V^* is canonically isomorphic to V (§1). Thus we have the

COROLLARY.

$$(V \otimes W)^* \cong V^* \otimes W^*.$$

The element F *of* $(V \otimes W)^*$ *corresponding to* $\varphi \otimes \psi \in V^* \otimes W^*$ *is given by* $F(v \otimes w) = \varphi(v)\psi(w)$ $(v \in V,\ w \in W)$.

PROOF.

$$(V \otimes W)^* = \mathrm{Hom}(V \otimes W, k) \cong \mathscr{L}(V, W; k) \cong V^* \otimes W^*.$$

Let us examine these isomorphisms. The second one is given by $V^* \otimes W^* \ni \varphi \otimes \psi \leftrightarrow \Phi \in \mathscr{L}(V, W; k)$, where $\Phi(v, w) = \varphi(v)\psi(w)$ (§4). The first isomorphism is given by $\mathscr{L}(V, W; k) \ni \Phi \leftrightarrow F \in (V \otimes W)^*$ such that $\Phi = F \circ \iota$. Thus, if F is the element corresponding to $\varphi \otimes \psi$, $F(v \otimes w) = F(\iota(v, w)) = \Phi(v, w) = \varphi(v)\psi(w)$. □

PROPOSITION 1.4. *By the correspondence* $(\alpha \otimes v \leftrightarrow \alpha v)$, *where* $\alpha \in K$, $v \in V$,

$$k \otimes V \cong V.$$

REMARK. The proposition means that the given correspondence extends to a linear isomorphism between the two vector spaces. Notice that the correspondence does not depend on the choice of bases. The same remark applies to Proposition 1.5.

PROOF. It is sufficient to show that the pair consisting of V and the bilinear mapping $\iota: k \times V \to V$, $\iota(\alpha, v) = \alpha v$ satisfies the conditions (T1) and (T2) (for $\mathscr{L}(k, V; U)$). It then follows that (V, ι) is a tensor product of k and V. Therefore, by Theorem 1.1(2), we have a linear isomorphism $F_0: k \otimes V \to V$ such that $F_0(\alpha \otimes v) = \iota(\alpha, v) = \alpha v$.

Clearly (V, ι) satisfies (T1). Consider (T2). For $\Phi \in \mathscr{L}(k, V; U)$, define a mapping $F: V \to U$ by $F(v) = \Phi(1, v)$ $(v \in V)$. Then $F \in \operatorname{Hom}(V, U)$ and we have $F \circ \iota = \Phi$, since

$$F \circ \iota(\alpha, v) = F(\alpha v) = \Phi(1, \alpha v) = \alpha(\Phi(1, v)) = \Phi(\alpha, v). \quad \square$$

PROPOSITION 1.5 (commutativity of the tensor product). *By the correspondence* $(v \otimes w \leftrightarrow w \otimes v)$, *we have* $V \otimes W \cong W \otimes V$.

PROOF. By (T2) (for $V \otimes W$), the bilinear mapping $V \times W \to W \otimes V$ $(v, w) \mapsto w \otimes v$ induces a linear mapping

$$F: V \otimes W \to W \otimes V, \qquad F(v \otimes w) = w \otimes v.$$

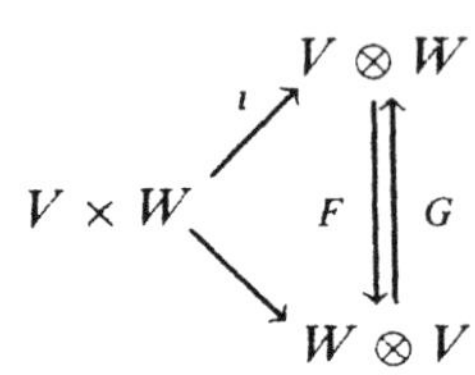

Similarly we have a linear mapping

$$G: W \otimes V \to V \otimes W, \qquad G(w \otimes v) = v \otimes w.$$

Since $F \circ G = \mathrm{id}$ and $G \circ F = \mathrm{id}$, F is an isomorphism. □

With respect to the tensor product of subspaces, we have the following result.

PROPOSITION 1.6 (tensor product of subspaces). *Let* V, W *be* k*-vector spaces and* V_1, W_1 *be their subspaces respectively. Let* $(V \otimes W, \iota)$ *be a tensor product of* V *and* W. *Let* $\iota_1 = \iota \,|\, V_1 \times W_1$ *be the restriction of* ι *to*

$V_1 \times W_1$ and U_1 be the subspace of $V \otimes W$ generated by $\iota(V_1 \times W_1)$. Then, the pair (U_1, ι_1) is a tensor product of V_1 and W_1.

PROOF. We will show that (U_1, ι_1) satisfies the conditions (T1) and (T2) for the tensor product of V_1 and W_1. (T1) follows from the definition. To prove (T2) choose $\Phi_1 \in \mathscr{L}(V_1, W_1; U)$. Then there exists $\Phi \in \mathscr{L}(V, W; U)$ such that $\Phi | V_1 \times W_1 = \Phi_1$. In fact, let $(e_1, \dots, e_{n_1})$ ($n_1 = \dim V_1$) and $(f_1, \dots, f_{m_1})$ ($m_1 = \dim W_1$) be bases of V_1 and W_1 respectively. Extend them to bases $(e_1, \dots, e_n)$, $(f_1, \dots, f_m)$ of V and W ($n = \dim V$, $m = \dim W$) and define $\Phi \in \mathscr{L}(V, W; U)$ by

$$\Phi(e_i, f_j) = \begin{cases} \Phi_1(e_i, f_j) & \text{if } 1 \le i \le n_1,\ 1 \le j \le m_1, \\ 0 & \text{otherwise.} \end{cases}$$

Then clearly $\Phi | V_1 \times W_1 = \Phi_1$. Φ induces a linear mapping $f: V \otimes W \to U$ such that $\Phi = F \circ \iota$. Let F_1 be the restriction of F to U_1. Then F_1 is a linear mapping of U_1 into U and we have, for every $(v, w) \in V_1 \times W_1$,

$$F_1 \circ \iota_1(v, w) = F \circ \iota(v, w) = \Phi(v, w) = \Phi_1(v, w),$$

which shows that $F_1 \circ \iota_1 = \Phi_1$. □

This proposition shows that the tensor product $V_1 \otimes W_1$ of subspaces can be identified naturally with a subspace of $V \otimes W$.

EXERCISE. Let V_1, V_2, V, W_1, W_2, and W be k-vector spaces. Show the existence of the following natural isomorphisms.

$$V \otimes (W_1 \oplus W_2) \cong (V \otimes W_1) \oplus (V \otimes W_2),$$
$$(V_1 \oplus V_2) \otimes W \cong (V_1 \otimes W) \oplus (V_2 \otimes W).$$

§6. Multilinear mappings and tensor products of more than two vector spaces

By increasing the number of variables of bilinear mappings, we obtain the concept of a multilinear mapping.

DEFINITION 1.3. Let $V_1, \dots, V_n$ and U be k-vector spaces. A mapping

$$\Phi: V_1 \times \dots \times V_n \to U$$

is called a *multilinear mapping* (or *n-multilinear mapping*) if Φ is linear in each variable. Namely, Φ is multilinear if we have, for each i $(1 \le i \le n)$, and for all $\alpha, \beta \in k$, $v_i, v_i' \in V_i$,

$$\Phi(v_1, \dots, \alpha v_i + \beta v_i', \dots, v_n) = \alpha\Phi(v_1, \dots, v_i, \dots, v_n) + \beta\Phi(v_1, \dots, v_i', \dots, v_n).$$

If U is the 1-dimensional space k, a multilinear mapping into k is called a *multilinear form* (or *n-multilinear form*). For example, let $\varphi_i \in V_i^*$, $1 \le i \le n$ and define $\Phi: V_1 \times \dots \times V_n \to k$ by $\Phi(v_1, \dots, v_n) = \varphi_1(v_1) \cdots \varphi_n(v_n)$. Then Φ is a multilinear form.

The set of multilinear mappings of $V_1 \times \cdots \times V_n$ into U is denoted by $\mathscr{L}(V_1, \ldots, V_n; U)$. If $n = 2$, we have a bilinear mapping, which we have already discussed. The properties (1), (2), and (3) of $\mathscr{L}(V, W; U)$ mentioned in §2 extend easily to $\mathscr{L}(V_1, \ldots, V_n; U)$. With respect to the linearization explained in §3, we have the following theorem.

THEOREM 1.1$'$. *Let $V_1, \ldots, V_n$ be k-vector spaces. There exist a k-vector space U_0 and an n-multilinear mapping $\iota \in \mathscr{L}(V_1, \ldots, V_n; U_0)$ that have the following properties* (T1) *and* (T2).

(T1) *U_0 is generated by the image $\iota(V_1 \times \cdots \times V_n)$ of ι.*

(T2) *For any $\Phi \in \mathscr{L}(V_1, \ldots, V_n; U)$, there exists a linear mapping F: $U_0 \to U$ such that $\Phi = F \circ \iota$.*

$$\begin{array}{ccc} V_1 \times \cdots \times V_n & \xrightarrow{\ \iota\ } & U_0 \\ & \searrow^{\Phi} & \downarrow F \\ & & U. \end{array}$$

The proof is similar to the case $n = 2$. Check it.

DEFINITION 1.4. The pair (U_0, ι) described in Theorem 1.1$'$ is called a *tensor product* of $V_1, \ldots, V_n$. We write $U_0 = V_1 \otimes \cdots \otimes V_n$ and

$$\iota(v_1, \ldots, v_n) = v_1 \otimes \cdots \otimes v_n \qquad (v_i \in V_i).$$

For $n = 3$, we have the following proposition.

PROPOSITION 1.7. *The correspondence $(v_1 \otimes v_2 \otimes v_3 \leftrightarrow (v_1 \otimes v_2) \otimes v_3)$ gives an isomorphism*

$$V_1 \otimes V_2 \otimes V_3 \cong (V_1 \otimes V_2) \otimes V_3.$$

PROOF. Consider the multilinear mapping $V_1 \times V_2 \times V_3 \to (V_1 \otimes V_2) \otimes V_3$ given by

$$(v_1, v_2, v_3) \mapsto (v_1 \otimes v_2) \otimes v_3.$$

From (T2) this mapping induces a linear mapping

$$F\colon V_1 \otimes V_2 \otimes V_3 \to (V_1 \otimes V_2) \otimes V_3, \qquad F(v_1 \otimes v_2 \otimes v_3) = (v_1 \otimes v_2) \otimes v_3.$$

Now, let us construct the inverse mapping of F. Fix an element $v \in V_3$ and consider the bilinear mapping $\Phi_v\colon V_1 \times V_2 \to V_1 \otimes V_2 \otimes V_3$ defined by $\Phi_v(v_1, v_2) = v_1 \otimes v_2 \otimes v$ $(v_1 \in V_1,\ v_2 \in V_2)$. By (T2) applied to $V_1 \otimes V_2$, Φ_v induces a linear mapping G_v:

$$G_v\colon V_1 \otimes V_2 \to V_1 \otimes V_2 \otimes V_3, \qquad G_v(v_1 \otimes v_2) = v_1 \otimes v_2 \otimes v.$$

$$\begin{array}{ccc} V_1 \times V_2 & \xrightarrow{\ \iota\ } & V_1 \otimes V_2 \\ & \searrow^{\Phi_v} & \downarrow G_v \\ & & V_1 \otimes V_2 \otimes V_3 \end{array} \qquad (v \in V_3).$$

For $v, v' \in V_3$, $\alpha \in k$, we have

$$G_{v+v'} = G_v + G_{v'}, \qquad G_{\alpha v} = \alpha G_v,$$

since G_v is uniquely defined by Φ_v. Using these facts, we define a bilinear mapping Ψ:

$$\Psi\colon (V_1 \otimes V_2) \times V_3 \to V_1 \otimes V_2 \otimes V_3,$$
$$\Psi(x, v) = G_v(x) \qquad (x \in V_1 \otimes V_2,\ v \in V_3).$$

$$\begin{array}{ccc} (V_1 \otimes V_2) \times V_3 & \xrightarrow{\ \iota\ } & (V_1 \otimes V_2) \otimes V_3 \\ & \searrow^{\Psi} & \downarrow G \\ & & V_1 \otimes V_2 \otimes V_3 . \end{array}$$

Then Ψ induces a linear mapping G:

$$G\colon (V_1 \otimes V_2) \otimes V_3 \to V_1 \otimes V_2 \otimes V_3, \qquad G(x \otimes v) = G_v(x).$$

By the definition, we have

$$G((v_1 \otimes v_2) \otimes v_3) = G_{v_3}(v_1 \otimes v_2) = v_1 \otimes v_2 \otimes v_3.$$

Thus $G \circ F =$ identity mapping of $V_1 \otimes V_2 \otimes V_3$, $F \circ G =$ identity mapping of $(V_1 \otimes V_2) \otimes V_3$. Thus, G is the inverse mapping of F and F is a linear isomorphism. □

Similarly,

COROLLARY. *The correspondence* $(v_1 \otimes v_2 \otimes v_3 \leftrightarrow v_1 \otimes (v_2 \otimes v_3))$ *gives an isomorphism*

$$V_1 \otimes V_2 \otimes V_3 \cong V_1 \otimes (V_2 \otimes V_3).$$

Combining these results, we have

PROPOSITION 1.8 (associativity of the tensor product). *The correspondence* $((v_1 \otimes v_2) \otimes v_3 \leftrightarrow v_1 \otimes (v_2 \otimes v_3))$ *gives an isomorphism*

$$(V_1 \otimes V_2) \otimes V_3 \cong V_1 \otimes (V_2 \otimes V_3).$$

In the following, we identify $V_1 \otimes V_2 \otimes V_3$, $(V_1 \otimes V_2) \otimes V_3$, and $V_1 \otimes (V_2 \otimes V_3)$ by these correspondences. In general, we can show by induction on n that, for any n vector spaces $V_1, V_2, \dots, V_n$, a tensor product of these vector spaces (in this order) is naturally isomorphic to $V_1 \otimes \cdots \otimes V_n$, no matter how parentheses are inserted. For example, in the case $n = 4$

$$\begin{aligned} ((V_1 \otimes V_2) \otimes V_3) \otimes V_4 &\cong (V_1 \otimes V_2) \otimes (V_3 \otimes V_4) \\ &\cong (V_1 \otimes (V_2 \otimes V_3)) \otimes V_4 \cong V_1 \otimes ((V_2 \otimes V_3) \otimes V_4) \\ &\cong V_1 \otimes (V_2 \otimes (V_3 \otimes V_4)) \cong (V_1 \otimes V_2 \otimes V_3) \otimes V_4 \\ &\cong V_1 \otimes (V_2 \otimes V_3 \otimes V_4) \cong V_1 \otimes V_2 \otimes V_3 \otimes V_4. \end{aligned}$$

We identify these spaces hereafter.

As in the case $n = 2$, we have the following results.

PROPOSITION 1.1′ (multilinearity of $\otimes$). *Let* $\alpha, \beta \in k$, $v_i, v_i' \in V_i$. *For any* i $(1 \le i \le n)$

$$\begin{aligned} &v_1 \otimes \cdots \otimes (\alpha v_i + \beta v_i') \otimes \cdots \otimes v_n \\ &\quad = \alpha(v_1 \otimes \cdots \otimes v_i \otimes \cdots \otimes v_n) + \beta(v_1 \otimes \cdots \otimes v_i' \otimes \cdots \otimes v_n). \end{aligned}$$

PROPOSITION 1.2′. *Let* $(e_1^{(i)}, \dots, e_{m_i}^{(i)})$ *be a basis for* V_i $(m_i = \dim V_i)$. *The* $m_1 \cdots m_n$ *elements* $e_{j_1}^{(1)} \otimes \cdots \otimes e_{j_n}^{(n)}$ $(1 \le j_i \le m_i,\ 1 \le i \le n)$ *form a basis of* $V_1 \otimes \cdots \otimes V_n$.

In particular,

$$\dim(V_1 \otimes \cdots \otimes V_n) = \dim V_1 \cdots \dim V_n.$$

PROPOSITION 1.3′.

$$V_1 \otimes \cdots \otimes V_n \cong \mathscr{L}(V_1^*, \dots, V_n^*; k).$$

§7. Tensor products of linear mappings

Let $F_1\colon V_1 \to W_1$ and $F_2\colon V_2 \to W_2$ be linear mappings, where V_i, W_i, $i = 1, 2$, are k-vector spaces. Then, as explained in the next theorem, F_1 and F_2 naturally induce a linear mapping of $V_1 \otimes V_2$ into $W_1 \otimes W_2$.

THEOREM 1.2. *Let* F_1 *and* F_2 *be mappings as above. Then there exists a unique linear mapping* $\widetilde{F}\colon V_1 \otimes V_2 \to W_1 \otimes W_2$ *such that*

$$\widetilde{F}(v_1 \otimes v_2) = F_1(v_1) \otimes F_2(v_2)$$

for all $v_1 \in V_1$, $v_2 \in V_2$. *The mapping* $\widetilde{F}$ *is called the tensor product of* F_1 *and* F_2 *and is denoted by* $F_1 \otimes F_2$.

PROOF. Let ι and κ be the canonical mappings of $V_1 \otimes V_2$ and $W_1 \otimes W_2$ respectively. Define a mapping $F_1 \times F_2\colon V_1 \times V_2 \to W_1 \times W_2$ as follows: for $(v_1, v_2) \in V_1 \times V_2$, $(F_1 \times F_2)(v_1, v_2) = (F_1(v_1), F_2(v_2))$.

$$\begin{array}{ccc} V_1 \times V_2 & \xrightarrow{\iota} & V_1 \otimes V_2 \\ {\scriptstyle F_1 \times F_2}\downarrow & & \downarrow{\scriptstyle \widetilde{F}} \\ W_1 \times W_2 & \xrightarrow[\kappa]{} & W_1 \otimes W_2 \end{array}$$

The composite mapping $\kappa \circ (F_1 \times F_2)$ of $F_1 \times F_2$ and κ is a bilinear mapping of $V_1 \times V_2$ into $W_1 \otimes W_2$. Therefore, by property (T) of $V_1 \otimes V_2$, there exists a unique linear mapping $\widetilde{F}\colon V_1 \otimes V_2 \to W_1 \otimes W_2$ such that $\kappa \circ (F_1 \times F_2) = \widetilde{F} \circ \iota$. This implies the relation above for $(v_1, v_2) \in V_1 \times V_2$. $\square$

The tensor product of linear mappings has the following properties.

PROPOSITION 1.9. *Let* V_i, W_i, *and* U_i $(i = 1, 2)$ *be* k*-vector spaces. For* $\alpha \in k$, $F_1, G_1 \in \mathrm{Hom}(V_1, W_1)$, $F_2, G_2 \in \mathrm{Hom}(V_2, W_2)$, *and* $H_j \in \mathrm{Hom}(W_j, U_j)$ $(j = 1, 2)$, *we have*

$$\begin{aligned} &F_1 \otimes (F_2 + G_2) = F_1 \otimes F_2 + F_1 \otimes G_2, \\ &(F_1 + G_1) \otimes F_2 = F_1 \otimes F_2 + G_1 \otimes F_2, \\ &(\alpha F_1) \otimes F_2 = F_1 \otimes (\alpha F_2) = \alpha(F_1 \otimes F_2), \\ &(H_1 \circ F_1) \otimes (H_2 \circ F_2) = (H_1 \otimes H_2) \circ (F_1 \otimes F_2). \end{aligned}$$

PROOF. As an illustration, we give the proof of the first equality. Since both sides of the equality are linear mappings of the vector space $V_1 \otimes V_2$

into $W_1 \otimes W_2$, it is enough to show that they coincide on the generating set $\{v_1 \otimes v_2 \mid v_1 \in V_1, v_2 \in V_2\}$ of $V_1 \otimes V_2$. For all $v_1 \otimes v_2$, we have

$$\begin{aligned}
&(F_1 \otimes (F_2 + G_2))(v_1 \otimes v_2) \\
&\quad = F_1(v_1) \otimes (F_2 + G_2)(v_2) \\
&\qquad\qquad \text{(property of the tensor product of mappings),} \\
&\quad = F_1(v_1) \otimes \{F_2(v_2) + G_2(v_2)\} \qquad \text{(definition of the sum of mappings),} \\
&\quad = F_1(v_1) \otimes F_2(v_2) + F_1(v_1) \otimes G_2(v_2) \qquad \text{(bilinearity of } \otimes\text{),} \\
&\quad = (F_1 \otimes F_2)(v_1 \otimes v_2) + (F_1 \otimes G_2)(v_1 \otimes v_2) \\
&\qquad\qquad \text{(property of the tensor product of mappings),} \\
&\quad = (F_1 \otimes F_2 + F_1 \otimes G_2)(v_1 \otimes v_2) \quad \text{(definition of the sum of mappings).} \quad \square
\end{aligned}$$

The *tensor product* $F_1 \otimes \cdots \otimes F_n : V_1 \otimes \cdots \otimes V_n \to W_1 \otimes \cdots \otimes W_n$ of n linear mappings $F_i : V_i \to W_i$ $(i = 1, \ldots, n)$ is defined similarly.

Next, we compute the matrix of $F_1 \otimes F_2$, using the matrices for F_1 and F_2 with respect to bases for V_i and W_i. Let (e_i), (e'_i), (f_j), and (f'_j) be bases of V_1, V_2, W_1, and W_2 respectively. Then $(e_i \otimes e'_j)$, $(f_i \otimes f'_j)$ are bases of $V_1 \otimes V_2$ and $W_1 \otimes W_2$ (Proposition 1.2). Let $A = (\alpha_{ij})$ and $B = (\beta_{ij})$ be the matrices for F_1 and F_2 with respect to the bases (e_i), (f_j), (e'_i), and (f'_j). Namely,

$$F_1(e_i) = \sum_l \alpha_{li} f_l, \qquad F_2(e'_j) = \sum_h \beta_{hj} f'_h.$$

Thus, for all i and j, we have

$$\begin{aligned}
(F_1 \otimes F_2)(e_i \otimes e'_j) &= F_1(e_i) \otimes F_2(e'_j) \\
&= \sum_l \sum_h \alpha_{li} \beta_{hj} (f_l \otimes f'_h).
\end{aligned}$$

When we arrange the bases $(e_i \otimes e'_j)$ and $(f_i \otimes f'_j)$ in lexicographical order (§5), the matrix of $F_1 \otimes F_2$ is given as follows:

$$\begin{aligned}
&(F_1 \otimes F_2)(e_1 \otimes e'_1, e_1 \otimes e'_2, \ldots) \\
&\quad = (f_1 \otimes f'_1, f_1 \otimes f'_2, \ldots) \begin{bmatrix} \alpha_{11}\beta_{11} & \alpha_{11}\beta_{12} & \cdots \\ \alpha_{11}\beta_{21} & \alpha_{11}\beta_{22} & \cdots \\ \vdots & \vdots & \end{bmatrix} \\
&\quad = (f_1 \otimes f'_1, f_1 \otimes f'_2, \ldots) \begin{bmatrix} \alpha_{11}B & \alpha_{12}B & \cdots & \alpha_{1n}B \\ \vdots & \vdots & & \vdots \\ \alpha_{m1}B & \cdots & \cdots & \alpha_{mn}B \end{bmatrix},
\end{aligned}$$

where $\dim V_1 = n$, $\dim W_1 = m$.

In general, we have the following definition.

DEFINITION 1.5. Let $A = (\alpha_{ij})$ and $B = (\beta_{ij})$ be matrices. The matrix

$$\begin{bmatrix} \alpha_{11}B & \alpha_{12}B & \cdots & \alpha_{1n}B \\ \alpha_{21}B & & & \\ \vdots & & & \vdots \\ \alpha_{m1}B & & \cdots & \alpha_{mn}B \end{bmatrix}$$

is called the *tensor product* (or *Kronecker product*) of A and B. It is denoted by $A \otimes B$. If A is an $m \times n$ matrix and B is an $m' \times n'$ matrix, $A \otimes B$ is an $mm' \times nn'$ matrix.

According to this definition, the matrix of $F_1 \otimes F_2$ with respect to the bases above is the tensor product of the matrices of F_1 and F_2.

EXAMPLE 1.3. Consider the case where $F_1(e_i) = \delta_{iq} f_p$, $F_2(e'_j) = \delta_{js} f'_r$. Then we have

$$(F_1 \otimes F_2)(e_i \otimes e'_j) = \delta_{iq}\delta_{js}(f_p \otimes f'_r)$$

(δ_{ij} = Kronecker's symbol). Namely, the tensor product of matrix units ([3]) E_{pq} and E_{rs} is given by

$$E_{pq} \otimes E_{rs} = E_{(p-1)m'+r,\,(q-1)n'+s},$$

where $n' = \dim V_2$, $m' = \dim W_2$.

The following formulas for the tensor product of matrices are obtained by rewriting Proposition 1.9 and Definition 1.5.

PROPOSITION 1.9′. *Let A_i be $m \times n$ matrices, B_i be $m' \times n'$ matrices; let C_1 be an $n \times l$ matrix and D_1 be an $n' \times l'$ matrix. Then we have*

$$\begin{aligned} &A_1 \otimes (B_1 + B_2) = A_1 \otimes B_1 + A_1 \otimes B_2, \\ &(A_1 + A_2) \otimes B_1 = A_1 \otimes B_1 + A_2 \otimes B_1, \\ &(\alpha A_1) \otimes B_1 = A_1 \otimes (\alpha B_1) = \alpha(A_1 \otimes B_1) \qquad (\alpha \in k), \\ &{}^t(A_1 \otimes B_1) = {}^tA_1 \otimes {}^tB_1, \\ &A_1C_1 \otimes B_1D_1 = (A_1 \otimes B_1)(C_1 \otimes D_1). \end{aligned}$$

COROLLARY. *If A and B are regular matrices, then $A \otimes B$ is regular, and we have*

$$(A \otimes B)^{-1} = A^{-1} \otimes B^{-1}.$$

This follows from the last formula in Proposition 1.9′.

It is an interesting question how certain properties of matrices A and B transfer to properties of $A \otimes B$. For instance,

PROPOSITION 1.10. *Let A be an $n \times n$ matrix whose eigenvalues are $\alpha_1, \ldots, \alpha_n$ and let B be an $m \times m$ matrix whose eigenvalues are $\beta_1, \ldots, \beta_m$. Then the eigenvalues of $A \otimes B$ are $\alpha_i\beta_j$ $(i = 1, \ldots, n,\ j = 1, \ldots, m)$.*

([3]) The matrix whose (i, j) component is 1 with all other components equal to 0 is denoted by E_{ij}. These matrices are called matrix units.

PROOF. Considering an extension field of k, if necessary, we may assume that A is similar to an upper triangular matrix with diagonal elements $\alpha_1, \ldots, \alpha_n$ and that B is similar to that of diagonal elements $\beta_1, \ldots, \beta_m$. Namely, there exist regular matrices S and T such that

$$S^{-1}AS = \begin{bmatrix} \alpha_1 & & * \\ & \ddots & \\ 0 & & \alpha_n \end{bmatrix}, \qquad T^{-1}BT = \begin{bmatrix} \beta_1 & & * \\ & \ddots & \\ 0 & & \beta_m \end{bmatrix}.$$

From the corollary to Proposition 1.9′, $S \otimes T$ is regular and $(S \otimes T)^{-1} = S^{-1} \otimes T^{-1}$. Hence

$$\begin{aligned}
(S \otimes T)^{-1}&(A \otimes B)(S \otimes T) \\
&= (S^{-1}AS) \otimes (T^{-1}BT) \\
&= \begin{bmatrix} \alpha & & * \\ & \ddots & \\ 0 & & \alpha_n \end{bmatrix} \otimes \begin{bmatrix} \beta_1 & & * \\ & \ddots & \\ 0 & & \beta_m \end{bmatrix} \\
&= \left[\begin{array}{c|c|c|c}
\begin{matrix} \alpha_1\beta_1 & & * \\ & \ddots & \\ & & \alpha_1\beta_m \end{matrix} & & & * \\ \hline
& \begin{matrix} \alpha_2\beta_1 & & * \\ & \ddots & \\ & & \alpha_2\beta_m \end{matrix} & & \\ \hline
& & \ddots & \\ \hline
0 & & & \begin{matrix} \alpha_n\beta_1 & & * \\ & \ddots & \\ & & \alpha_n\beta_m \end{matrix}
\end{array}\right],
\end{aligned}$$

which proves the proposition. □

COROLLARY. *Let A be an $n \times n$ matrix and B an $m \times m$ matrix. The trace* $\operatorname{tr}(A \times B)$ *and the determinant* $\det(A \otimes B)$ *of* $A \otimes B$ *are given by*

$$\operatorname{tr}(A \otimes B) = \operatorname{tr}(A) \cdot \operatorname{tr}(B),$$
$$\det(A \otimes B) = (\det(A))^m \cdot (\det(B))^n.$$

PROOF. Let X be a square matrix whose eigenvalues are $\eta_1, \ldots, \eta_n$. Then

$$\operatorname{tr}(X) = \eta_1 + \cdots + \eta_n,$$
$$\det(X) = \eta_1\eta_2 \cdots \eta_n.$$

The results follow easily from these formulas. □

EXERCISE 1. Let A and B be matrices. Assume that $\operatorname{rank}(A) = r_1$ and $\operatorname{rank}(B) = r_2$. Compute the $\operatorname{rank}(A \otimes B)$ of $A \otimes B$.

EXERCISE 2. Let $A_1, \dots, A_n$ be square matrices. Prove that, if $A_1 \otimes \cdots \otimes A_n$ is equal to a scalar matrix αI_N, then each A_i is a scalar matrix and $\alpha = \alpha_1 \cdots \alpha_n$, where $A_i = \alpha_i I_{n_i}$.

§8. Examples of tensor product spaces

(a) The space of linear mappings. Let V and W be k-vector spaces. The k-vector space $\mathrm{Hom}(V, W)$ consisting of all linear mappings of V into W can be regarded as a tensor product space.

Let $\varphi \in V^*$, $w \in W$. For $(\varphi, w) \in V^* \times W$, we define an element $F_{\varphi, w} \in \mathrm{Hom}(V, W)$ as follows:

$$F_{\varphi, w}(v) = \varphi(v)w \qquad (v \in V).$$

It is easy to see that $F_{\varphi, w}$ is an element of $\mathrm{Hom}(V, W)$.

PROPOSITION 1.11. *The correspondence* $(\varphi \otimes w \leftrightarrow F_{\varphi, w})$, *gives an isomorphism between* $V^* \otimes W$ *and* $\mathrm{Hom}(V, W)$.

PROOF. The mapping $\Phi\colon V^* \times W \to \mathrm{Hom}(V, W)$, which assigns $F_{\varphi, w}$ to $(\varphi, w) \in V^* \times W$, is clearly bilinear. Therefore, by condition (T2) applied to $V^* \otimes W$, there exists a linear mapping $\widetilde{F}\colon V^* \otimes W \to \mathrm{Hom}(V, W)$ such that $\widetilde{F} \circ \iota = \Phi$.

$$\begin{array}{ccc} V^* \times W & \xrightarrow{\ \iota\ } & V^* \otimes W \\ & \searrow^{\Phi} & \downarrow \widetilde{F} \\ & & \mathrm{Hom}(V, W) \end{array}$$

It suffices to show that $\widetilde{F}$ is an isomorphism. First, we will prove that $\widetilde{F}$ is one-to-one. Since $\widetilde{F}$ is linear, it suffices to deduce $t = 0$ from $\widetilde{F}(t) = 0$ for $t \in V^* \otimes W$. Suppose $\widetilde{F}(t) = 0$ for $t \in V^* \otimes W$. t can be written in the form $t = \sum_{i=1}^r (\varphi_i \otimes w_i)$, where $\varphi_i \in V^*$, $w_i \in W$. We can assume that $\{w_1, \dots, w_r\}$ is linearly independent (cf. §5). Then,

$$\begin{aligned} \widetilde{F}(t) &= \widetilde{F}\left(\sum_{i=1}^r (\varphi_i \otimes w_i)\right) = \widetilde{F}\left(\sum_{i=1}^r \iota(\varphi_i, w_i)\right) \\ &= \sum_i (\widetilde{F} \circ \iota)(\varphi_i, w_i) = \sum_i \Phi(\varphi_i, w_i) \\ &= \sum_i F_{\varphi_i, w_i}. \end{aligned}$$

$\widetilde{F}(t) = 0$ implies $\widetilde{F}(t)(v) = \sum_i F_{\varphi_i, w_i}(v) = \sum_i \varphi_i(v) w_i = 0$ for all $v \in V$. Since $\{w_1, \dots, w_r\}$ is linearly independent, $\varphi_i(v) = 0$, $i = 1, \dots, r$, for all $v \in V$, which implies $\varphi_i = 0$, namely, $t = 0$.

To show that $\widetilde{F}$ is surjective set $\dim V = n$ and $\dim W = m$. Take a basis (e_i) for V, the dual basis (e_i^*) for (e_i), and a basis (f_j) for W. Then

$$F_{e_i^*, f_j}(e_k) = e_i^*(e_k) f_j = \delta_{ik} f_j.$$

Therefore the $F_{e_i^*, f_j}$ $(1 \leq i \leq n, 1 \leq j \leq m)$ form a basis for $\mathrm{Hom}(V, W)$. On the other hand, we know that $F_{e_i^*, f_j} = \Phi(e_i^*, f_j) = \widetilde{F}(e_i^* \otimes f_j)$, which implies that $\widetilde{F}$ is surjective. □

Note that we proved both the injectivity and surjectivity of $\widetilde{F}$. But, since $\dim \mathrm{Hom}(V, W) = \dim(V^* \otimes W)$, each of these facts implies the other since $\widetilde{F}$ is linear. Generalizing this observation, we have

LEMMA 1.2. *Let V_1, V_2, and U be k-vector spaces and let $\tilde{\imath}: V_1 \times V_2 \to U$ be a bilinear mapping. Assume that they satisfy the following conditions:*

(1) $\dim U = \dim V_1 \cdot \dim V_2$,

(2) *the image of $\tilde{\imath}$ generates U.*

Then, $(U, \tilde{\imath})$ is a tensor product of V_1 and V_2. Namely, there exists a linear isomorphism $F: V_1 \otimes V_2 \to U$ such that $F(v_1 \otimes v_2) = \tilde{\imath}(v_1, v_2)$.

$$\begin{array}{ccc} V_1 \times V_2 & \xrightarrow{\ \imath\ } & U_0 = V_1 \otimes V_2 \\ & \searrow^{\tilde{\imath}} & \downarrow F \\ & & U \end{array}$$

PROOF. Set $U_0 = V_1 \otimes V_2$ and $\imath(v_1, v_2) = v_1 \otimes v_2$. By the property (T2) of tensor product $V_1 \otimes V_2$, the bilinear mapping $\tilde{\imath}$ induces a linear mapping $F: U_0 \to U$ such that $\tilde{\imath} = F \circ \imath$. Since the image of $\tilde{\imath}$ is contained in that of F, by condition (2) above, the image of F is equal to U, i.e., F is surjective. Condition (1) implies that $\dim U_0 = \dim U$. Thus F is an isomorphism and, by construction, $\tilde{\imath} = F \circ \imath$.

COROLLARY TO PROPOSITION 1.11. *If $W = V$ in Proposition 1.11, we have*

$$\mathrm{Hom}(V, V) \cong V^* \otimes V.$$

With regard to Lemma 1.2 we make a remark about the tensor product of linear mappings.

Let $F_i \in \mathrm{Hom}(V_i, W_i)$, $i = 1, 2$. Then $F_1 \otimes F_2 \in \mathrm{Hom}(V_1 \otimes V_2, W_1 \otimes W_2)$ is defined in §7, and we have a bilinear mapping:

$$\begin{array}{ccc} \mathrm{Hom}(V_1, W_1) \times \mathrm{Hom}(V_2, W_2) & \to & \mathrm{Hom}(V_1 \otimes V_2, W_1 \otimes W_2). \\ \Psi & & \Psi \\ (F_1, F_2) & \mapsto & F_1 \otimes F_2 \end{array}$$

Using Lemma 1.2, we have

$$\mathrm{Hom}(V_1, W_1) \otimes \mathrm{Hom}(V_2, W_2) \cong \mathrm{Hom}(V_1 \otimes V_2, W_1 \otimes W_2).$$

Here, condition (2) is verified by the computation in Example 1.3. Thus we can regard the tensor product of linear mappings defined in §7 as a tensor product of k-vector spaces $\mathrm{Hom}(V_i, W_i)$, $i = 1, 2$. In particular, we can apply the results on the tensor product of vector spaces to the tensor product of linear mappings and of matrices.

Consider the case where $W_1 = W_2 = k$ in the above formula. Since $W_1 \otimes W_2 = k$, we get again

$$V_1^* \otimes V_2^* \cong (V_1 \otimes V_2)^*,$$

which is the corollary to Proposition 1.3.

(b) Extension of the field of scalars. Let V be a k-vector space and let k_0 be a subfield of k. Then V can be regarded naturally as a vector space over k_0. This procedure is called *restriction of the field of scalars* of V to k_0. The k_0-vector space V is denoted by V_{k_0}. Note that the dimension of V_{k_0} over k_0 is given by $\dim_{k_0} V_{k_0} = nd$, where $d = [k : k_0] =$ the degree of the extension $=$ the dimension of k as vector space over k_0 and $n = \dim_k V =$ the dimension of V over k. Now, let k_1 be an extension field of k, i.e., k_1 is a field which contains k as subfield. Then, using the tensor product, we can naturally construct a vector space over k_1 from V.

The field k_1 can be regarded as a vector space over k, and we assume that $[k_1 : k]$ is finite since we have been discussing only finite-dimensional vector spaces. But, in fact this assumption is not necessary as we explain later (§9). An example of such a k_1 and k is given by $k_1 = \mathbf{C}$ (the field of complex numbers) and $k = \mathbf{R}$ (the field of real numbers), where $[k_1 : k] = 2$.

Now, construct a tensor product $k_1 \otimes V$, where k_1 is regarded as a k-vector space. Then $k_1 \otimes V$ is a k-vector space. In particular, addition is defined. For an element ξ of k_1 and an element x of $k_1 \otimes V$, we shall define the scalar multiplication ξx as follows.

For $\xi \in k_1$, the mapping

$$\Phi_\xi : k_1 \times V \to k_1 \otimes V, \qquad \Phi_\xi(\alpha, v) = \xi\alpha \otimes v$$

is obviously bilinear. For example, $\Phi_\xi(\delta\alpha, v) = \xi\delta\alpha \otimes v = \delta\xi\alpha \otimes v = \delta\Phi_\xi(\alpha, v)$, $\Phi_\xi(\alpha, \delta v) = \xi\alpha \otimes \delta v = \delta\Phi_\xi(\alpha, v)$ for $\delta \in k$. From property (T) of the tensor product, Φ_ξ induces a unique k-linear mapping

$$L_\xi : k_1 \otimes V \to k_1 \otimes V$$

such that $L_\xi \circ \iota = \Phi_\xi$ (i.e., $L_\xi(\alpha \otimes v) = \xi\alpha \otimes v$). For $\xi \in k_1$ and $x \in k_1 \otimes V$, define ξx by $\xi x = L_\xi(x)$. We have to verify that this operation satisfies the conditions for scalar multiplication.

Since L_ξ is linear, for $x, y \in k_1 \otimes V$, $L_\xi(x + y) = L_\xi(x) + L_\xi(y)$, which implies $\xi(x + y) = \xi x + \xi y$. Let $\xi, \xi' \in k_1$. Then

$$L_\xi \circ L_{\xi'} = L_{\xi\xi'}, \qquad L_\xi + L_{\xi'} = L_{\xi+\xi'}, \qquad L_1 = \mathrm{id}.$$

These formulas can be verified using the uniqueness of L_ξ for Φ_ξ. Rewriting the formulas, we have, for any $x \in k_1 \otimes V$,

$$\xi(\xi' x) = (\xi\xi')x,$$

$$\xi x + \xi' x = (\xi + \xi')x, \qquad 1 \cdot x = x.$$

For $\xi \in k$, $L_\xi(x)$ is equal to ξx which is initially defined in $k_1 \otimes V$.

Thus $k_1 \otimes V$ can be considered as vector space over k_1, which is called the *extension of the field of scalars* of V to k_1 and is denoted by V^{k_1}.

Then we have

PROPOSITION 1.12. (1) *The mapping* $F_0\colon V \to (V^{k_1})_k$ (= *the restriction of the field of scalars of* V^{k_1} *to* k) *which assigns* $1 \otimes v \in (V^{k_1})_k$ *to* $v \in V$ *is a one-to-one* k*-linear mapping.*

(2) V^{k_1} *is generated by* $F_0(V)$ *as a vector space over* k_1. *In particular, if* $(e_1, \dots, e_n)$ *is a* (k-) *basis of the* k*-vector space* V, $(1 \otimes e_1, \dots, 1 \otimes e_n)$ *is a* (k_1-) *basis of the* k_1*-vector space* V^{k_1}. *Thus we have* $\dim_{k_1} V^{k_1} = \dim_k V$.

PROOF. (1) Note that $(V^{k_1})_k$ is nothing but $k_1 \otimes V$ as a k-vector space. For $v, v' \in V$, $\xi \in k$, clearly $F_0(v + v') = F_0(v) + F_0(v')$ and $F_0(\xi v) = 1 \otimes \xi v = \xi \otimes v = \xi(1 \otimes v) = \xi F_0(v)$, which shows that F_0 is a k-linear mapping.

Let $v \in \operatorname{Ker} F_0$ (= the kernel of F_0). Namely, $F_0(v) = 1 \otimes v = 0$, which implies $v = 0$ (cf. Corollary to Proposition 1.2). Thus F_0 is one-to-one.

(2) Let $x = \sum_i(\xi_i \otimes v_i) \in V^{k_1} = k_1 \otimes V$ $(\xi_i \in k_1, v_i \in V)$. Then

$$\sum_i(\xi_i \otimes v_i) = \sum_i \xi_i(1 \otimes v_i) = \sum \xi_i F_0(v_i).$$

This implies that V^{k_1} is generated by $F_0(V)$. Any element of $F_0(V)$ is a linear combination of $1 \otimes e_1, \dots, 1 \otimes e_n$ with coefficients in k. And $F_0(V)$ generates the k_1-vector space V^{k_1}. Therefore we know that $1 \otimes e_1, \dots, 1 \otimes e_n$ generate V^{k_1}.

Next, we will show that these elements are linearly independent over k_1. Suppose that

$$\sum_j \xi_j(1 \otimes e_j) = 0,$$

where $\xi_j \in k_1$. Then

$$0 = \sum_j \xi_j(1 \otimes e_j) = \sum_j(\xi_j \otimes e_j).$$

Here, $\sum(\xi_j \otimes e_j)$ is an element of the k-vector space $k_1 \otimes V$ and $\{e_1, \dots, e_n\}$ is linearly independent over k. Therefore, by Lemma 1.1 each $\xi_j = 0$. □

Identifying V and the image of F_0 in V^{k_1}, we can regard as $V \subseteq V^{k_1}$.

We also define the extension of linear mappings.

PROPOSITION 1.13. *Let* V, W *be* k*-vector spaces and* F *a* k*-linear mapping of* V *into* W. *Then there exists a unique* k_1*-linear mapping* $\widetilde{F}$ *of* V^{k_1} *into* W^{k_1} *such that the restriction* $\widetilde{F} \mid V$ *of* $\widetilde{F}$ *to* V *is equal to* F, *where we regard as* $V^{k_1} \supseteq V$ *and* $W^{k_1} \supseteq W$.

PROOF. Let id be the identity mapping of k_1, which we can consider as a k-linear mapping of the k-vector space k_1 into itself. By taking the tensor product of linear mappings, we have a k-linear mapping $\mathrm{id} \otimes F\colon k_1 \otimes V \to k_1 \otimes W$. We will prove that $\mathrm{id} \otimes F$ is a k_1-linear mapping when $k_1 \otimes V$

and $k_1 \otimes W$ are regarded as k_1-vector spaces V^{k_1} and W^{k_1} respectively. It suffices to show $(\mathrm{id} \otimes F)(\xi x) = \xi((\mathrm{id} \otimes F)(x))$ for all $\xi \in k_1$, $x \in k_1 \otimes V$, namely, for all $\xi \in k_1$,

$$(\mathrm{id} \otimes F) \circ L_\xi = L_\xi \circ (\mathrm{id} \otimes F).$$

(The L_ξ on the left side is a mapping of $k_1 \otimes V$ and the L_ξ on the right side is a mapping of $k_1 \otimes W$.) Now, for $\alpha \otimes v \in k_1 \otimes V$, we have

$$((\mathrm{id} \otimes F) \circ L_\xi)(\alpha \otimes v) = (\mathrm{id} \otimes F)(\xi\alpha \otimes v) = \xi\alpha \otimes F(v),$$
$$(L_\xi \circ (\mathrm{id} \otimes F))(\alpha \otimes v) = L_\xi(\alpha \otimes F(v)) = \xi\alpha \otimes F(v).$$

Since the set $\{\alpha \otimes v \mid \alpha \in k_1,\ v \in V\}$ generates $k_1 \otimes V$ as a k-vector space, these formulas show that k-linear mappings $(\mathrm{id} \otimes F) \circ L_\xi$ and $L_\xi \circ (\mathrm{id} \otimes F)$ coincide on $k_1 \otimes V$. Thus $\widetilde{F} = \mathrm{id} \otimes F$ is a k_1-linear mapping and satisfies the condition.

For the uniqueness, suppose that the k_1-linear mappings $\widetilde{F}$ and $\widetilde{F}'$ satisfy the condition $\widetilde{F} \mid V = \widetilde{F}' \mid V$, i.e., $\widetilde{F} \mid F_0(V) = \widetilde{F}' \mid F_0(V)$. Since $F_0(V)$ generates the k_1-vector space V^{k_1}, and both $\widetilde{F}$ and $\widetilde{F}'$ are k_1-linear, it follows that $\widetilde{F} = \widetilde{F}'$. □

Definition 1.6. The k_1-linear mapping $\widetilde{F}$ obtained from Proposition 1.13 is called the extension of F to k_1 and is denoted by F^{k_1}.

Let (e_i) and (f_j) be bases for V and W respectively. Then $(1 \otimes e_i)$ and $(1 \otimes f_j)$ are bases for V^{k_1} and W^{k_1}. The matrix of F^{k_1} with respect to these bases is the same as that for F with respect to (e_i) and (f_j). This follows easily from the construction.

With regard to extensions of mappings we have the following more general result.

Proposition 1.14. *Let V be a k-vector space, W a k_1-vector space, and G a k-linear mapping of V into W_k. Then there exists a unique k_1-linear mapping $\widetilde{G}$ of V^{k_1} into W such that $\widetilde{G} \circ F_0 = G$ (where $\widetilde{G}$ is regarded as a k-linear mapping).*

$$\begin{array}{ccc} V & \xrightarrow{F_0} & V^{k_1} \\ {\scriptstyle G}\downarrow & & \downarrow{\scriptstyle \widetilde{G}} \\ W_k & = & W \end{array}$$

Proof. Define a mapping $\Phi: k_1 \times V \to W_k$ which assigns $\xi G(v) \in W_k$ to $(\xi, v) \in k_1 \times V$. Obviously Φ is bilinear. Therefore by property (T2) of the tensor product $k_1 \otimes V$, Φ induces a k-linear mapping $\widetilde{G}: k_1 \otimes V \to W_k$ such that $\widetilde{G}(\xi \otimes v) = \Phi(\xi, v) = \xi G(v)$.

$$\begin{array}{ccc} k_1 \times V & \xrightarrow{\iota} & k_1 \otimes V \\ & \overset{\Phi}{\searrow} & \downarrow{\scriptstyle \widetilde{G}} \\ & & W_k \end{array}$$

We will prove that $\widetilde{G}$ is a k_1-linear mapping. For $\xi' \in k_1$

$$\begin{aligned}(\widetilde{G} \circ L_{\xi'})(\xi \otimes v) &= \widetilde{G}(\xi'(\xi \otimes v)) = \widetilde{G}(\xi'\xi \otimes v) = \xi'\xi G(v) \\ &= \xi'\widetilde{G}(\xi \otimes v) \\ &= (L_{\xi'} \circ \widetilde{G})(\xi \otimes v) \qquad (\xi \otimes v \in k_1 \otimes V).\end{aligned}$$

By the same argument as for Proposition 1.13, we have for all $x \in k_1 \otimes V$,

$$\widetilde{G}(\xi' x) = \xi' \widetilde{G}(v).$$

The uniqueness of $\widetilde{G}$ is obvious. □

Using this proposition we can prove the following theorem.

THEOREM 1.3. *Let $\widetilde{V}$ be a k_1-vector space and let F_1 be a k-linear mapping of V into $\widetilde{V}_k$. Assume they satisfy the following conditions* (E1) *and* (E2).

(E1) *$\widetilde{V}$ is generated by $F_1(V)$ as a k_1-vector space.*

(E2) *For any k_1-vector space W and any k-linear mapping $G \in$* Hom(V, W_k), *there exists a k_1-linear mapping $\widetilde{G} \in$* Hom$(\widetilde{V}, W)$ *such that $\widetilde{G} \circ F_1 = G$, where $\widetilde{G}$ is regarded as a k-linear mapping.*

Then there exists a k_1-linear isomorphism κ of V^{k_1} onto $\widetilde{V}$ such that $\kappa \circ F_0 = F_1$, where κ is regarded as a k-linear mapping.

PROOF. By (E1) we know that the $\widetilde{G}$ given in (E2) is unique. By Proposition 1.14, there is a k_1-linear mapping $\widetilde{F}_1 : V^{k_1} \to \widetilde{V}$ such that $\widetilde{F}_1 \circ F_0 = F_1$.

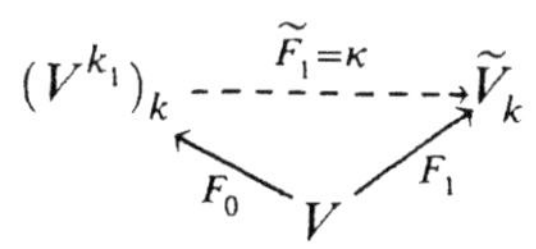

It suffices to show that $\kappa = \widetilde{F}_1$ is an isomorphism. Given condition (E2) and the uniqueness of $\widetilde{G}$ in (E2), we can use the same argument as in the proof of the uniqueness of the tensor product (cf. §3, Theorem 1.1). □

This theorem *characterizes the extension* V^{k_1} of the field of scalars. In other words, for a k-vector space V and an extension field k_1, we can define without ambiguity the extension V^{k_1} as a pair $(\widetilde{V}, F_1)$ consisting of a k_1-vector space $\widetilde{V}$ and k-linear mapping $F_1 : V \to \widetilde{V}_k$ satisfying the conditions (E1) and (E2).

EXAMPLE 1.4. Let V be a vector space over $\mathbf{R}$ and $V^{\mathbf{C}}$ be the extension of the field of scalars of V to $\mathbf{C}$. Choose a basis $(e_1, \dots, e_n)$ of V and a basis $(1, \sqrt{-1})$ of the vector space $\mathbf{C}$ over $\mathbf{R}$. Every element of $V^{\mathbf{C}}$ can be uniquely expressed as a linear combination of $1 \otimes e_i$ and $\sqrt{-1} \otimes e_i$ $(1 \le i \le n)$ with real coefficients.

$$x = \sum_i \alpha_i (1 \otimes e_i) + \sum_i \beta_i \left(\sqrt{-1} \otimes e_i\right) \qquad (\alpha_i, \beta_i \in \mathbf{R}).$$

Since $F_0(\sum_i \alpha_i e_i) = \sum_i \alpha_i(1 \otimes e_i)$, we can write $\sum_i \alpha_i(1 \otimes e_i) = \sum_i \alpha_i e_i$, identifying $V \subseteq V^{\mathbf{C}}$. Since

$$\sqrt{-1} \otimes e_i = L_{\sqrt{-1}}(1 \otimes e_i) = \sqrt{-1}(1 \otimes e_i),$$

we can also write $\sum_i \beta_i(\sqrt{-1} \otimes e_i) = \sqrt{-1}(\sum_i \beta_i(1 \otimes e_i)) = \sqrt{-1}(\sum_i \beta_i e_i)$. Thus every element in $V^{\mathbf{C}}$ admits an expression of the form $v_1 + \sqrt{-1}v_2$, where $v_i \in V$. It is easy to see that this expression is unique.

For $x = v_1 + \sqrt{-1}v_2$ ($v_i \in V$), define $\bar{x} = v_1 - \sqrt{-1}v_2$. Then we have $\overline{x + x'} = \bar{x} + \bar{x'}$, $\overline{\xi x} = \bar{\xi}\bar{x}$, $\bar{\bar{x}} = x$, for $x, x' \in V^{\mathbf{C}}$ and $\xi \in \mathbf{C}$.

§9. Construction of tensor products with generators and relations

In this section we shall examine a third method of constructing tensor products. Though this method is not so easy to understand, it is theoretically interesting, since it can be widely applied.

First, we describe a *k-vector space generated by a set* A, where A is a nonempty arbitrary (infinite or finite) set. Consider a k-vector space $\mathscr{V}(A)$ which has a basis $\{e(a) \mid a \in A\}$ such that, between the elements of A and the basis elements, there is a bijective correspondence $e(a) \leftrightarrow a$ $(a \in A)$. $\dim \mathscr{V}(A)$ is equal to the cardinality of A, therefore, it is not necessarily finite. $\mathscr{V}(A)$ is called a vector space generated by A. Namely, $\mathscr{V}(A)$ is the set of all finite linear combinations of $e(a)$ $(a \in A)$ with coefficients in k,

$$\sum_{a \in A} \alpha(a)e(a) \qquad (\alpha(a) \in k).$$

Here, "finite" means that the number of elements $a \in A$ with $\alpha(a) \neq 0$ is finite. Two linear combinations $\sum_{a \in A} \alpha(a)e(a)$ and $\sum_{a \in A} \beta(a)e(a)$ are equal if and only if $\alpha(a) = \beta(a)$ for every $a \in A$. Addition and the scalar multiplication are defined as follows:

$$\sum_{a \in A} \alpha(a)e(a) + \sum_{a \in A} \beta(a)e(a) = \sum_{a \in A} (\alpha(a) + \beta(a))e(a),$$

$$\gamma \cdot \sum_{a \in A} \alpha(a)e(a) = \sum_{a \in A} (\gamma\alpha(a))e(a) \qquad (\gamma \in k).$$

There is another way to define $\mathscr{V}(A)$. The set $\mathscr{F}$ of all k-valued functions defined on A is a k-vector space with respect to usual additional and scalar multiplication. The subset consisting of all $f \in \mathscr{F}$ such that $f(a) = 0$ except for a finite number of elements of A is a subspace. From the correspondence

$$\mathscr{V}(A) \ni \sum_{a \in A} \alpha(a)e(a) \mapsto f, \quad \text{where } f(a) = \alpha(a) \ (a \in A),$$

it follows that $\mathscr{V}(A)$ is isomorphic to this subspace as a k-vector space.

Consider the mapping $e\colon A \to \mathscr{V}(A)$ which assigns $e(a_0)$ to $a_0 \in A$, where $e(a_0)$ is a basis element $(= \sum_{a\in A}\alpha(a)e(a)$, $\alpha(a_0) = 1$, and $\alpha(a) = 0$ for $a \neq a_0)$. Then the pair $(\mathscr{V}(A), e)$ has the following *universality property.*

PROPOSITION 1.15. *Let W be a k-vector space and ρ a mapping of A into W. Then there exists a unique k-linear mapping F of $\mathscr{V}(A)$ into W such that $F \circ e = \rho$.*

PROOF. Define a mapping F of $\mathscr{V}(A)$ into W by $F(\sum_{a\in A}\alpha(a)e(a)) = \sum_{a\in A}\alpha(a)\rho(a)$. Obviously F is linear and $F\circ e = \rho$. □

Now, let us construct a tensor product of k-vector spaces V and W. Take the direct product set $V\times W (= \{(v, w) \mid v \in V, w \in W\})$ as A in the discussion above and construct a k-vector space $\mathscr{V}(V \times W)$. To simplify the notation we denote $e((v, w))$ by $e(v, w)$ for $(v, w) \in V\times W$ and the terms with $\alpha(v, w) = 0$ in a sum $\sum \alpha(v, w)e(v, w)$ are omitted. We also write $1\cdot e(v, w) = e(v, w)$.

Consider the subspace X of $\mathscr{V}(V \times W)$ generated by the following elements:

$$\begin{cases} e(v_1 + v_2, w) - e(v_1, w) - e(v_2, w), \\ e(v, w_1 + w_2) - e(v, w_1) - e(v, w_2), \\ e(\alpha v, w) - \alpha e(v, w), \\ e(v, \alpha w) - \alpha e(v, w) \qquad (v_1, v_2, v \in V,\ w_1, w_2, w \in W,\ \alpha\in k). \end{cases}$$

In order to obtain bilinearity of the product $\otimes$, we consider these elements.

Denote by U_1 the factor space of $\mathscr{V}(V\times W)$ by X. Let π be the canonical mapping of $\mathscr{V}(V\times W)$ onto U_1. From the definition of X, it follows that, for instance,

$$\pi(e(v_1 + v_2, w)) - \pi(e(v_1, w)) - \pi(e(v_2, w)) = 0.$$

Let $\iota_1 = \pi\circ e$, where $e\colon V\times W \to \mathscr{V}(V\times W)$ is given by $(v, w)\mapsto e(v, w)$.

THEOREM 1.4. *The pair (U_1, ι_1) constructed above is a tensor product of V and W.*

PROOF. By the definition of X, ι_1 is a bilinear mapping of $V\times W$ into U. We shall show that the pair (U_1, ι_1) satisfies the conditions (T1) and (T2) in Theorem 1.1. Since $\{e(v, w) \mid (v, w) \in V\times W\}$ generates $\mathscr{V}(V\times W)$, U_1 is generated by $\{\iota_1(v, w) \mid (v, w)\in V\times W\}$, which proves (T1).

To prove (T2), let Φ be a bilinear mapping of $V\times W$ into a vector space U. By Proposition 1.15, there is a linear mapping $\tilde{F}\colon \mathscr{V}(V\times W) \to U$ such that $\tilde{F}(e(v, w)) = \Phi(v, w)$ $((v, w)\in V\times W)$.

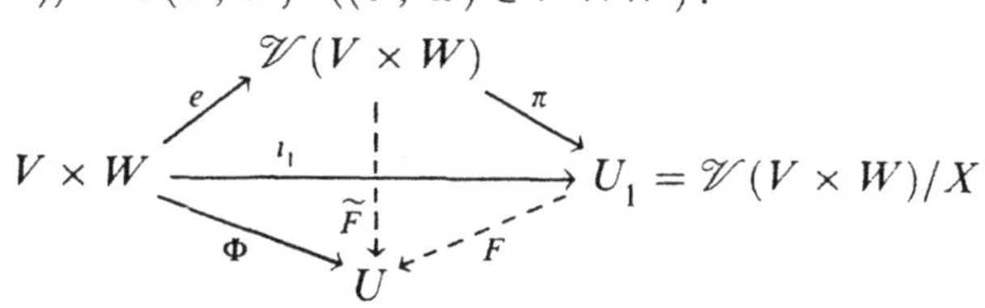

Since Φ is bilinear, all generators of X are contained in the kernel $\operatorname{Ker}\tilde{F}$ of $\tilde{F}$. For example,

$$\begin{aligned}
&\tilde{F}(e(v_1+v_2, w) - e(v_1, w) - e(v_2, w)) \\
&\quad = \tilde{F}(e(v_1+v_2, w)) - \tilde{F}(e(v_1, w)) - \tilde{F}(e(v_2, w)) \quad \text{(linearlity of } \tilde{F}) \\
&\quad = \Phi(v_1+v_2, w) - \Phi(v_1, w) - \Phi(v_2, w) \\
&\quad = 0 \qquad\qquad\qquad\qquad\qquad\qquad\qquad\qquad \text{(bilinearlity of } \Phi).
\end{aligned}$$

Since $\operatorname{Ker}\tilde{F}$ is a subspace, X is contained in $\operatorname{Ker}\tilde{F}$. Therefore $\tilde{F}$ induces a linear mapping F of the factor space $U_1 = \mathscr{V}(V \times W)/X$ into U such that $\tilde{F} = F \circ \pi$. Then we have $F \circ \iota_1 = F \circ \pi \circ e = \tilde{F} \circ e = \Phi$. □

REMARK. In the construction above of (U_1, ι_1), it is not necessary to assume that V and W are finite dimensional. Thus, we can construct a k-vector space satisfying (T1) and (T2) for arbitrary k-vector spaces V and W.

In particular, in the discussion of §8(b) Extension of the field of scalars, the arguments there are still valid without the assumption that $[k_1 : k]$ is finite. Thus, for any extension field k_1 of k, we can construct V^{k_1}.

NOTE. Let V be a k-vector space (not necessarily finite dimensional) and let W be a subspace of V. Then the cosets $v + W$ of W in V $(v \in V)$ form a k-vector space, which is called the *factor space* V/W of V by W. Addition and scalar multiplication in V/W are defined as follows: for $v, v' \in V$, $\alpha \in k$,

$$(v + W) + (v' + W) = (v + v') + W, \qquad \alpha(v + W) = \alpha v + W.$$

The mapping π of V onto V/W defined by $\pi(v) = v + W$ is linear and is called the canonical mapping. The pair $(V/W, \pi)$ has the following universal property: Let U be a vector space and f a linear mapping of V into U such that $\operatorname{Ker} f$ contains W. Then there exists a unique linear mapping $\tilde{f}: V/W \to U$ such that $f = \tilde{f} \circ \pi$.

§10. Tensor products of R-modules

Modules over a ring are a generalization of the concept of vector spaces. The definition of tensor product generalizes and we present a theorem generalizing Theorem 1.1. In this section we shall discuss this generalization. These results can be easily obtained as a variation of the construction explained in §9. We do not go into the details, since the further properties of a tensor product of modules generally depend on the properties of rings. First we give several definitions.

DEFINITION 1.7. Let P be a set. P is called a *module* if there is a rule which associates with every (ordered) pair m_1, m_2 of elements of P a third element $m_1 + m_2$ of P, which is called the sum of m_1 and m_2, and if the following conditions are satisfied:

For all $m_1, m_2, m_3 \in P$,

$$(m_1 + m_2) + m_3 = m_1 + (m_2 + m_3) \quad \text{(associativity)}.$$

There exists an element $0 \in P$ such that, for all $m \in P$,

$$m + 0 = 0 + m = m.$$

For each $m \in P$, there exists an element $m' \in P$ such that $m + m' = m' + m = 0$. (m' is denoted by $-m$.)

For all $m_1, m_2 \in P$,

$$m_1 + m_2 = m_2 + m_1 \quad \text{(commutativity)}.$$

DEFINITION 1.8. Let R be a set. R is called a *ring* if there are two rules which associate with each ordered pair r_1, r_2 of R, their sum $r_1 + r_2 \in R$ and product $r_1 r_2 \in R$ and if the following conditions are satisfied:

R is a module with respect to addition.

For all $r_1, r_2, r_3 \in R$,

$$(r_1 r_2) r_3 = r_1 (r_2 r_3) \qquad \text{(associativity)},$$

$$\left.\begin{aligned} (r_1 + r_2) r_3 &= r_1 r_3 + r_2 r_3, \\ r_1 (r_2 + r_3) &= r_1 r_2 + r_1 r_3. \end{aligned}\right\} \quad \text{(distributivity)}.$$

Let R be a ring. If there is an element 1 such that, for all $r \in R$,

$$r1 = 1r = r,$$

then the element 1 is called the *unit element* of R.

REMARK. Precisely speaking, what we call rules in the definitions are mappings of product sets. For example, the sum in a module P is a mapping $s: P \times P \to P$ and we denote the image $s(m_1, m_2)$ $((m_1, m_2) \in P \times P)$ by $m_1 + m_2$. Similarly, the sum and the product in a ring R are also mappings of $R \times R$ into R. The conditions on these operations can be expressed as conditions on these mappings. For instance, associativity in a module P is expressed as "for all $m_1, m_2, m_3 \in P$, we have $s(s(m_1, m_2), m_3) = s(m_1, s(m_2, m_3))$" and commutativity is "for all $m_1, m_2 \in P$, we have $s(m_1, m_2) = s(m_2, m_1)$." In a ring R, let s and m be the mappings which give the sum and product respectively. Then associativity is expressed in a similar fashion and the distributivity condition is "for all $r_1, r_2, r_3 \in R$, we have $m(s(r_1, r_2), r_3) = s(m(r_1, r_3), m(r_2, r_3))$ and $m(r_1, s(r_2, r_3)) = s(m(r_1, r_2), m(r_1, r_3))$".

In this section R denotes an arbitrary ring with the unit element 1. As a generalization of vector space we define left R-modules and right R-modules.

DEFINITION 1.9. Let M be a module. M is called a *left R-module* if there is a rule which associates with every $r \in R$ and $m \in M$, an element rm of M and if the following conditions are satisfied: (in other words, there exists a mapping $R \times M \to M$, $(r, m) \mapsto rm$ such that:)

For all $r_1, r_2, r \in R$, $m_1, m_2, m \in M$,

$$r(m_1 + m_2) = rm_1 + rm_2, \qquad (r_1 + r_2)m = r_1 m + r_2 m,$$
$$(r_1 r_2)m = r_1(r_2 m), \qquad 1m = m.$$

Similarly, a module M is called a *right R-module* if there is a mapping $M \times R \to M$, $(m, r) \to mr$ such that:

For all $r_1, r_2, r \in R$, $m_1, m_2, m \in M$,

$$(m_1 + m_2)r = m_1 r + m_2 r,$$
$$m(r_1 + r_2) = mr_1 + mr_2,$$
$$m(r_1 r_2) = (mr_1)r_2,$$
$$m1 = m.$$

The difference between a left R-module and a right one consists in the order of operation of the product element $r_1 r_2 \in R$ on $m \in M$. In the case of a left module, the above condition shows that r_2 operates first then r_1 operates. For a right module, the order is reversed. To show this concretely, we write r on the left or the right of m.

EXAMPLE 1.5. If R is a field k, a left k-module (or a right k-module) is a k-vector space.

EXAMPLE 1.6. If a ring R is commutative (namely, if we have $r_1 r_2 = r_2 r_1$ for all $r_1, r_2 \in R$), it is not necessary to distinguish between left and right R-modules. In this case, we call them simply R-modules. (cf. Example 1.5.)

In fact, let M be a left R-module. Define a mapping $M \times R \to R$ by $mr = rm$. (The right side has meaning, since M is a left R-module.) Then, with this mapping, M becomes a right R-module. For instance,

$$\begin{aligned} m(r_1 r_2) &\underset{\text{definition}}{=} (r_1 r_2)m \underset{\text{commutativity of } R}{=} (r_2 r_1)m \\ &\underset{M \text{ is a left } R\text{-module}}{=} r_2(r_1 m) \underset{\text{definition}}{=} r_2(mr_1) \\ &\underset{\text{definition}}{=} (mr_1)r_2, \end{aligned}$$

EXAMPLE 1.7. Let $R = \mathbf{Z}$ (the ring of integers) and let M be a module. For $n \in \mathbf{N}$ (the set of natural numbers), define

$$nm = m + \cdots + m \quad (n \text{ terms}), \quad (-n)m = -(nm), \quad 0m = 0.$$

From this definition, it follows that any module can be regarded as a $\mathbf{Z}$-module.

As a generalization of a basis for a vector space, we will define a free module and a basis for it. Here, we describe the case of a left R-module M, but all steps can be applied to right modules.

A subset $\{m_i\}$ of M is called a *set of generators* (or a generating set) if every element of M can be expressed as a linear combination of a finite number of $\{m_i\}$ with coefficients in R. In this case, we also say that M is generated by $\{m_i\}$. A subset $\{m_i\}$ of M is *R-free* if there do not exist nontrivial linear relations with coefficients in R, namely, whenever we have a linear relation, for $m_{i_j} \in \{m_i\}$

$$r_1 m_{i_1} + \cdots + r_l m_{i_l} = 0 \qquad (r_j \in R),$$

then $r_j = 0$ for all j.

A R-free set of generators of M is called an *R-basis* of M and an R-module M which has an R-basis is called a *free R-module.* As we know, a k-vector space (i.e., when R is a field k) always has bases. But, for an arbitrary ring, that is not the case.

As a generalization of bilinear mapping, we define the concept of a balanced mapping.

DEFINITION 1.10. Let M be a right R-module, N a left R-module, and P a module. A mapping $\Phi\colon M \times N \to P$ is called an *R-balanced mapping* if the following conditions are satisfied: for all $m_1, m_2, m \in M$, $n_1, n_2, n \in N$, $r \in R$, we have

$$\Phi(m_1 + m_2, n) = \Phi(m_1, n) + \Phi(m_2, n),$$
$$\Phi(m, n_1 + n_2) = \Phi(m, n_1) + \Phi(m, n_2),$$
$$\Phi(m, rn) = \Phi(mr, n).$$

The existence theorem for tensor products (Theorem 1.1) generalizes as follows.

THEOREM 1.5. *Let M be a right R-module and N a left R-module, then*

(1) *There exist a module P_0 and an R-balanced mapping $\iota\colon M \times N \to P_0$ which satisfy the following conditions* (T1) *and* (T2).

(T1) *P_0 is generated by $\iota(M \times N)$, where P_0 is regarded as a $\mathbf{Z}$-module.*

(T2) *For every R-balanced mapping $\Phi\colon M \times N \to P$, there exists a module homomorphism $F\colon P_0 \to P$ such that $\Phi = F \circ \iota$.*

(2) *The pair (P_0, ι) is unique in the following sense: If the pairs (P_0, ι) and (P_0', ι') consisting of a module and an R-balanced mapping satisfy the conditions* (T1) *and* (T2), *then there exists a unique module isomorphism $F_0\colon P_0 \to P_0'$ such that $F_0 \circ \iota = \iota'$.*

This theorem can be proved in a way similar to the case of k-vector spaces. Part (2) is proved by the argument in §3. For Part (1), we can use the method described in §9.

Let X be a free $\mathbf{Z}$-module which has a $\mathbf{Z}$-basis $(e(m, n))$ where $e(m, n)$ corresponds bijectively to $(m, n) \in M \times N$. Let Y be the $\mathbf{Z}$-submodule of X generated by the following elements:

$$\begin{cases} e(m_1 + m_2, n) - e(m_1, n) - e(m_2, n), \\ e(m, n_1 + n_2) - e(m, n_1) - e(m, n_2), \\ e(m, rn) - e(mr, n) \qquad (m, m_1, m_2 \in M,\ n, n_1, n_2 \in N,\ r \in R). \end{cases}$$

Then consider the factor module P_0 of X by Y and the canonical homomorphism $\pi\colon X \to P_0$ which assigns $x + Y$ to $x \in X$. Define $\iota\colon M \times N \to P_0$

by $\iota(m, n) = \pi(e(m, n))$. We can prove that the pair (P_0, ι) has the required properties.

NOTE. Let P and P' be modules. A mapping $F: P \to P'$ is called a *module homomorphism* if $F(x+y) = F(x) + F(y)$ for all $x, y \in P$. If a module homomorphism F is one-to-one and onto, F is called a module isomorphism.

A nonempty subset P' of a module P is called a *submodule* of P if we have $x + y \in P'$ and $-x \in P'$ whenever x and y are elements of P'.

For a subset S of P, the intersection of all submodules of P containing S is also a submodule which contains S. It is called the submodule generated by S.

Let P be a module and P' a submodule of P. We can define a *factor module* P/P' in a way similar to the case of vector spaces.

DEFINITION 1.11. The pair (P_0, ι) (or P_0) is called a *tensor product* of M and N. We denote P_0 by $M \otimes_R N$ and $\iota(m, n) = m \otimes n$.

Let S be a ring, and assume that M is not only a right R-module but also a left S-module. Moreover, we assume that the operations of R and S on M are commutative [4], namely, we have, for all $m \in M$, $s \in S$, $r \in R$,

$$(sm)r = s(mr).$$

Then we can prove that $M \otimes_R N$ is a left S-module. As in the case of extension of the field of scalars (§8(b)), for a fixed element s of S, define a mapping

$$\begin{array}{ccc} M \times N & \to & M \otimes_R N \\ \cup\!\!\!\shortmid & & \cup\!\!\!\shortmid \\ (m, n) & \mapsto & sm \otimes n \end{array}$$

This mapping is an R-balanced mapping. (To check it, we use the commutativity of R and S on M.) Thus there exists a module homomorphism of $M \otimes_R N$,

$$L_s: M \otimes_R N \to M \otimes_R N, \qquad L_s(m \otimes n) = sm \otimes n,$$

and we can define $sx = L_s(x)$ for $x \in M \otimes_R N$. By this operation, $M \otimes_R N$ is a left S-module.

In particular, when R is a field k, since k is commutative, a k-vector space is a (k, k)-bimodule and the tensor product of k-vector spaces is a special case of the discussion here.

Exercises

In exercises 1 and 2, let $V = \mathbf{R}^3$ and $W = \mathbf{R}^2$ be the real vector spaces consisting of all the column vectors of dimension 3 and 2 respectively. Take the basis $\mathscr{E}$ of V and $\mathscr{F}$ of W consisting of the unit vectors as follows.

---

(4) Then M is called an (S, R)-*bimodule*.

$$\mathscr{E} = \left\{ e_1 = \begin{bmatrix} 1 \\ 0 \\ 0 \end{bmatrix}, e_2 = \begin{bmatrix} 0 \\ 1 \\ 0 \end{bmatrix}, e_3 = \begin{bmatrix} 0 \\ 0 \\ 1 \end{bmatrix} \right\},$$

$$\mathscr{F} = \left\{ f_1 = \begin{bmatrix} 1 \\ 0 \end{bmatrix}, f_2 = \begin{bmatrix} 0 \\ 1 \end{bmatrix} \right\}.$$

1. Let x be the element of $V \otimes W$ given by

$$x = \begin{bmatrix} -1 \\ 2 \\ 3 \end{bmatrix} \otimes \begin{bmatrix} 1 \\ -2 \end{bmatrix}.$$

(1) Express x as a linear combination of the basis elements $(e_i \otimes f_j)$.

(2) Define e_i' $(i = 1, 2, 3)$ and f_j' $(j = 1, 2)$ as follows.

$$e_1' = e_1 - e_2, \quad e_2' = e_2 - e_3, \quad e_3' = e_1 + e_2 + e_3,$$
$$f_1' = f_1 - f_2, \quad f_2' = f_1 + f_2.$$

Compute the matrix of the change of bases from $(e_i \otimes f_j)$ to $(e_i' \otimes f_j')$.

(3) Compute the components of x with respect to the basis $(e_i' \otimes f_j')$.

(4) Is it possible to express the following element y as a sum with 1 or 2 terms of the form $v \otimes w$ $(v \in V,\ w \in W)$?

$$y = \begin{bmatrix} 1 \\ 2 \\ 1 \end{bmatrix} \otimes \begin{bmatrix} 1 \\ 2 \end{bmatrix} + \begin{bmatrix} 2 \\ 1 \\ 1 \end{bmatrix} \otimes \begin{bmatrix} 2 \\ 1 \end{bmatrix} - \begin{bmatrix} 3 \\ 3 \\ 2 \end{bmatrix} \otimes \begin{bmatrix} 3 \\ 3 \end{bmatrix}.$$

NOTE. Let X be a k-vector space. Let $\mathscr{G} = (x_i)$ and $\mathscr{G}' = (x_i')$ be bases of X. The matrix (p_{ij}) given by

$$x_j' = \sum_i p_{ij} x_i$$

is called the matrix of the change of bases from (x_i) to (x_i').

2. (1) If we assign $(\varphi(e_1), \varphi(e_2), \varphi(e_3))$ to $\varphi \in V^*$, we have a mapping T of V^* into the real vector space consisting of all 3-dimensional row vectors of real coefficients. Verify that T is a linear isomorphism. Compute the row vectors corresponding to the dual basis elements of $\mathscr{E}' = (e_1', e_2', e_3')$ defined in 1.(2), by T.

Using T, we will identify V^* with the space of 3-dimensional row vectors.

(2) Use the isomorphism $V^* \otimes W \cong \mathrm{Hom}(V, W)$, given by Proposition 1.11, to compute the matrix of the linear mapping corresponding to

$$(3, 2, 4) \otimes \begin{bmatrix} 2 \\ 1 \end{bmatrix}$$

with respect to the bases $\mathscr{E}$ and $\mathscr{F}$.

(3) By the same isomorphism as above, compute the element of $V^* \otimes W$ corresponding to the linear mapping whose matrix with respect to the bases $\mathscr{E}$ and $\mathscr{F}$ is

$$\begin{pmatrix} 1 & 2 & 3 \\ 2 & 3 & 4 \end{pmatrix}.$$

In the following exercises, V and W denote k-vector spaces.

3. Let $\mathscr{E} = (e_i)$, $\widetilde{\mathscr{E}} = (\tilde{e}_i)$ be bases for V and let (α_{ij}) be the matrix of the change of bases from $\mathscr{E}$ to $\widetilde{\mathscr{E}}$. Also, let $\mathscr{F} = (f_i)$, $\widetilde{\mathscr{F}} = (f_i)$ be bases of W and let (β_{pq}) be the matrix of the change of bases from $\mathscr{F}$ to $\widetilde{\mathscr{F}}$. Namely, we have

$$\tilde{e}_j = \sum_i \alpha_{ij} e_i \qquad (j = 1, \dots, \dim V),$$

$$\tilde{f}_q = \sum_p \beta_{pq} f_p \qquad (q = 1, \dots, \dim W).$$

Compute the matrix of the change of bases from $(e_i \otimes f_p)$ to $(e_i' \otimes f_p')$, where the basis elements are arranged in lexicographical order.

4. Let u be a nonzero element of $V \otimes W$. Write u in the form $u = \sum_{i=1}^r (v_i \otimes w_i)$ $(v_i \in V,\ w_i \in W)$. Show that, if $\{v_1, \dots, v_r\}$ or $\{w_1, \dots,$ $w_r\}$ is linearly dependent, we can decrease the number of terms in the expression, i.e., we have another expression $u = \sum_{i=1}^{r'} (v_i' \otimes w_i')$ $(v_i' \in V,$ $w_i' \in W,\ r' < r)$. Therefore, if, in the expression $u = \sum_{i=1}^r (v_i \otimes w_i)$, the number r of terms is minimal, $\{v_1, \dots, v_r\}$ and $\{w_1, \dots, w_r\}$ are linearly independent.

5. By the isomorphism $\operatorname{Hom}(V, W) \cong V^* \otimes W$ given by Proposition 1.11, let t_F be an element of $V^* \otimes W$ corresponding to $F \in \operatorname{Hom}(V, W)$. Show that, if $\operatorname{rank} F$ is equal to r, t_F can be written in the form $t_F = \sum_{i=1}^r (\varphi_i \otimes w_i)$ $(\varphi_i \in V^*,\ w_i \in W)$ with r terms, and t_F cannot be expressed as a sum of less than r terms.

6 (Tensor product of factor spaces). Let V_1 (resp. W_1) be a subspace of V (resp. W). By Proposition 1.6, $V_1 \otimes W$ and $V \otimes W_1$ can be regarded as subspaces of $V \otimes W$. Let X be the subspace of $V \otimes W$ generated by $V_1 \otimes W$ and $V \otimes W_1$. Let U_0 be the factor space $(V \otimes W)/X_0$ and $\pi: V \otimes W \to U_0$ the canonical mapping

(1) Prove that, for $(v + V_1, w + W_1) \in (V/V_1) \times (W/W_1)$, the element $\pi(v \otimes w)$ is independent of the choice of coset representatives v and w, and that, by this correspondence, we can define a bilinear mapping $\iota_0: (V/V_1) \times (W/W_1) \to U_0$.

(2) Prove that (U_0, ι_0) is a tensor product of (V/V_1) and (W/W_1).

7. Let $F \in \operatorname{Hom}(V \otimes W, V \otimes W)$. Assume that $F \circ (\mathrm{id}_V \otimes G) = (\mathrm{id}_V \otimes G) \circ F$ for any $G \in \operatorname{Hom}(W, W)$. Prove that there exists an element $F_0 \in \operatorname{Hom}(V, V)$ such that $F = F_0 \otimes \mathrm{id}_W$, where id_V (resp. id_W) is the identity mapping of V (resp. W).

CHAPTER II

Tensors and Tensor Algebras

In this chapter we define tensors. Tensors are elements of vector spaces that are obtained by taking tensor products of several copies of a k-vector space V and of its dual space V^*. When we define tensors this way, we can show that their components satisfy the transformation law, which gives the classical definition of tensors. Next, we define symmetric tensors and alternating (skew symmetric) tensors, and discuss their properties.

The tensor algebra, which possesses a certain universality, is also defined.

§1. Definition and examples of tensors

Let V be a k-vector space and let V^* be the dual space of V. Let T be a vector space which is obtained as the tensor product of several copies of V and of V^*, for example $(V \otimes V) \otimes V^*$ or $V \otimes V^* \otimes V^* \otimes V$. Then T is generally called a *tensor space.*

From the associativity of the tensor product (Proposition 1.8) we can, when we make a tensor space, neglect parentheses. Therefore a tensor space is isomorphic to some $V^{\varepsilon_1} \otimes \cdots \otimes V^{\varepsilon_r}$, the tensor product of a sequence $V^{\varepsilon_1}, \ldots, V^{\varepsilon_r}$ consisting of V and V^*, where we set $V = V^1$, $V^* = V^{-1}$, and $\varepsilon_i = \pm 1$. Elements of this space are called tensors of type $(\varepsilon_1, \ldots, \varepsilon_r)$.

We can make a further identification. Suppose that T is isomorphic to $V^{\varepsilon_1} \otimes \cdots \otimes V^{\varepsilon_r}$ and T' is isomorphic to $V^{\varepsilon'_1} \otimes \cdots \otimes V^{\varepsilon'_r}$. (Both have the same number of factors.) If the number of V among $V^{\varepsilon_1}, \ldots, V^{\varepsilon_r}$ and that among $V^{\varepsilon'_1}, \ldots, V^{\varepsilon'_r}$ are equal (so the numbers of V^* are also equal), then there is a natural isomorphism between T and T' obtained from the commutativity of the tensor product (Proposition 1.5, $V \otimes V^* \cong V^* \otimes V$), and we can identify T and T' (for example, $T = V \otimes V \otimes V^* \otimes V$ and $T' = V^* \otimes V \otimes V \otimes V$). Therefore, to discuss a tensor space T, the numbers of V and V^* factors are important. If T is a tensor product of p copies of V and q copies of V^*, we call T the tensor space of type (p, q) and denote it by $T_q^p(V)$. Thus, if we identify as above,

$$T_q^p(V) = \underbrace{V \otimes \cdots \otimes V}_{p \text{ factors}} \otimes \underbrace{V^* \otimes \cdots \otimes V^*}_{q \text{ factors}}.$$

Set $T_0^{\ 0}(V) = k$. Sometimes we write $T_0^{\ p}(V) = T^p(V)$, $T_q^{\ 0}(V) = T_q(V)$, and $T_0^{\ 0}(V) = T^0(V) = T_0(V)$.

Elements of the tensor space $T_q^{\ p}(V)$ are called *tensors of type* (p, q) or tensors which are *contravariant of degree* (or order) p *and covariant of degree* (or order) q. In particular, tensors of type $(p, 0)$ are called contravariant tensors of degree p and those of type $(0, q)$ covariant tensors of degree q. Moreover the elements of $T_0^{\ 1}(V) = V$ are called contravariant vectors, those of $T_1^{\ 0}(V) = V^*$ covariant vectors, and those of $T_0^{\ 0}(V) = k$ scalars.

We will give several examples of tensors, using the isomorphisms presented in Chapter 1.

EXAMPLE 2.1. Since $\mathrm{Hom}(V, V) \cong V^* \otimes V \cong V \otimes V^*$ (see the corollary to Proposition 1.11),

$$\mathrm{Hom}(V, V) \cong T_1^{\ 1}(V).$$

That is, the *linear transformations of* V can be regarded as *tensors of type* $(1, 1)$. In this case, to $\varphi \otimes v \in V^* \otimes V$ $(\varphi \in V^*,\ v \in V)$, there corresponds the element $T_{\varphi, v} \in \mathrm{Hom}(V, V)$ defined by the following formula

$$T_{\varphi, v}(x) = \varphi(x)v \qquad (x \in V).$$

And this correspondence extends to the entire space.

EXAMPLE 2.2. Since $V \otimes W \cong \mathscr{L}(V^*, W^*; k)$ (I, §4) and $(V^*)^* \cong V$, we have

$$\mathscr{L}(V, V; k) \cong V^* \otimes V^* = T_2^{\ 0}(V).$$

(This result can be obtained also by Proposition 1.3 and its corollary.) That is, the *bilinear forms* on V are *covariant tensors of degree* 2.

In a similar fashion to the case of 2 variables, we generally obtain

$$\mathscr{L}\underbrace{(V, \dots, V}_{q \text{ factors}}; k) \cong \underbrace{V^* \otimes \cdots \otimes V^*}_{q \text{ factors}} = T_q^{\ 0}(V).$$

In this case $\varphi_1 \otimes \cdots \otimes \varphi_q$ $(\varphi_i \in V^*) \in V^* \otimes \cdots \otimes V^*$ is regarded as an element of $\mathscr{L}(V, \dots, V; k)$ as follows:

$$(\varphi_1 \otimes \cdots \otimes \varphi_q)(v_1 \otimes \cdots \otimes v_q) = \varphi_1(v_1) \cdots \varphi_q(v_q) \qquad (v_i \in V).$$

More generally, we have

$$T_q^{\ p}(V) \cong \mathscr{L}\underbrace{(V, \dots, V}_{q \text{ factors}}, \underbrace{V^*, \dots, V^*}_{p \text{ factors}}; k).$$

In fact,

$$\begin{aligned} T_q^{\ p}(V) &= \underbrace{V^* \otimes \cdots \otimes V^*}_{q \text{ factors}} \otimes \underbrace{V \otimes \cdots \otimes V}_{p \text{ factors}} \\ &\cong V^* \otimes \cdots \otimes V^* \otimes (V^*)^* \otimes \cdots \otimes (V^*)^* \\ &\cong \mathscr{L}(V, \dots, V, V^*, \dots, V^*; k). \end{aligned}$$

$t = \psi_1 \otimes \cdots \otimes \psi_q \otimes w_1 \otimes \cdots \otimes w_p \in T_q^{\ p}(V)$ $(\psi_i \in V^*,\ w_j \in V)$ is regarded as an element of $\mathscr{L}(V, \ldots, V, V^*, \ldots, V^*; k)$ by

$$\begin{aligned} &t(v_1, \ldots, v_q, \varphi_1, \ldots, \varphi_p) \\ &\quad = \varphi_1(w_1) \cdots \varphi_p(w_p)\psi_1(v_1) \cdots \psi_q(v_q) \qquad (v_i \in V,\ \varphi_j \in V^*). \end{aligned}$$

By this isomorphism, we can consider an element of $T_q^{\ p}(V)$ as a $(p+q)$-*multilinear form* defined on the product of q *copies of* V *and* p *copies of* V^*.

EXAMPLE 2.3. $(T_q^{\ p}(V))^* \cong T_p^{\ q}(V)$. That is, the dual space of $T_q^{\ p}(V)$ is $T_p^{\ q}(V)$.

In fact, since

$$T_q^{\ p}(V) = \underbrace{V \otimes \cdots \otimes V}_{p \text{ factors}} \otimes \underbrace{V^* \otimes \cdots \otimes V^*}_{q \text{ factors}},$$

by the corollary to Proposition 1.3, we have

$$(T_q^{\ p}(V))^* = \underbrace{V^* \otimes \cdots \otimes V^*}_{p \text{ factors}} \otimes \underbrace{V \otimes \cdots \otimes V}_{q \text{ factors}} = T_p^{\ q}(V).$$

Here $t = \psi_1 \otimes \cdots \otimes \psi_p \otimes w_1 \otimes \cdots \otimes w_q \in T_p^{\ q}(V)$, $(\psi_i \in V^*, w_j \in V)$ is regarded as a function on $T_q^{\ p}(V)$ by

$$\begin{aligned} &t(\varphi_1 \otimes \cdots \otimes \varphi_q \otimes v_1 \otimes \cdots \otimes v_p) \\ &\quad = \varphi_1(w_1) \cdots \varphi_q(w_q)\psi_1(v_1) \cdots \psi_p(v_p) \in k \qquad (\varphi_i \in V^*,\ v_j \in V). \end{aligned}$$

We can verify this easily, examining the correspondence of each isomorphism between spaces.

EXAMPLE 2.4 (Bilinear multiplication). Setting $W = U = V$ in the following formula

$$\mathscr{L}(V, W; U) \cong \operatorname{Hom}(V \otimes W, U) \cong (V \otimes W)^* \otimes U \cong V^* \otimes W^* \otimes U,$$

we have

$$\mathscr{L}(V, V; V) \cong V^* \otimes V^* \otimes V = T_2^{\ 1}(V).$$

Here the element m of $\mathscr{L}(V, V; V)$ corresponding to $\varphi_1 \otimes \varphi_2 \otimes v$ $(\varphi_1, \varphi_2 \in V^*,\ v \in V)$ by these isomorphisms, is given by

$$m(v_1, v_2) = \varphi_1(v_1)\varphi_2(v_2)v \qquad (v_1, v_2 \in V).$$

In fact, let us examine the isomorphisms above. To $\varphi_1 \otimes \varphi_2 \otimes v$ there corresponds $\Phi \otimes v \in (V \otimes V)^* \otimes V$, where $\Phi \in (V \otimes V)^*$ is such that $\Phi(v_1 \otimes v_2) = \varphi_1(v_1)\varphi_2(v_2)$. To $\Phi \otimes v$ there corresponds $T \in \operatorname{Hom}(V \otimes V, V)$ such that $T(x) = \Phi(x)v$ $(x \in V \otimes V)$. Finally we have

$$m(v_1, v_2) = T(v_1 \otimes v_2) = \Phi(v_1 \otimes v_2)v = \varphi_1(v_1)\varphi_2(v_2)v.$$

Let m be a bilinear mapping $V \times V \to V$. If we call $m(v_1, v_2) \in V$ $(v_1, v_2 \in V)$ the product of v_1 and v_2, we can define a multiplication of V. As usual, we denote $m(v_1, v_2) = v_1 v_2$. By definition we have

$$\begin{aligned}
&(v_1 + v_1')v_2 = v_1 v_2 + v_1' v_2, \\
&v_1(v_2 + v_2') = v_1 v_2 + v_1 v_2' \quad \text{(distributivity)}, \\
&(\alpha v_1)v_2 = v_1(\alpha v_2) = \alpha(v_1 v_2) \\
&\qquad (v_1, v_1', v_2, v_2' \in V, \ \alpha \in k).
\end{aligned}$$

Thus, an element of $T_2^{\ 1}(V)$ defines a bilinear multiplication of V and vice versa. Since defining a multiplication on V is equivalent to giving an element $m \in T_2^{\ 1}(V)$, the properties of a multiplication can be described in terms of m (cf. 1, §10, Remark).

§2. Properties of tensor spaces

(a) Components of tensors and their transformation laws. Let V be a k-vector space and let V^* be the dual space of V. In this section we describe the properties of the tensor space

$$T_q^{\ p}(V) = \underbrace{V \otimes \cdots \otimes V}_{p \text{ factors}} \otimes \underbrace{V^* \otimes \cdots \otimes V^*}_{q \text{ factors}}$$

defined in §1.

If $\dim_k V = n$, $\dim_k T_q^{\ p}(V) = n^{p+q}$ since the dimension of a tensor product is the product of the dimensions of the factors. Choose a basis $\mathscr{E} = (e_1, \ldots, e_n)$ of V. Let $(f^1, \ldots, f^n)$ be the dual basis of $\mathscr{E}$. Then, the following n^{p+q} elements

$$\{e_{i_1} \otimes \cdots \otimes e_{i_p} \otimes f^{j_1} \otimes \cdots \otimes f^{j_q} | 1 \le i_\nu \le n, \ 1 \le j_\mu \le n\}$$

form a basis for $T_q^{\ p}(V)$ (Proposition 1.2′). We call this basis the *standard basis* for $T_q^{\ p}(V)$ corresponding to $\mathscr{E}$. To simplify the notation, set

$$e_{i_1} \otimes \cdots \otimes e_{i_p} \otimes f^{j_1} \otimes \cdots \otimes f^{j_q} = t_{i_1, \ldots, i_p}^{\quad j_1, \ldots, j_q}.$$

Then an element z of $T_q^{\ p}(V)$ can be uniquely expressed as

$$z = \sum_{i_1, \ldots, i_p, j_1, \ldots, j_q} \xi_{j_1, \ldots, j_q}^{\quad i_1, \ldots, i_p} t_{i_1, \ldots, i_p}^{\quad j_1, \ldots, j_q} \qquad (\xi_{j_1, \ldots, j_q}^{\quad i_1, \ldots, i_p} \in k).$$

The n^{p+q} elements $\{\xi_{j_1, \ldots, j_q}^{\quad i_1, \ldots, i_p}\}$ $(1 \le i_\nu \le n, \ 1 \le j_\mu \le n)$ of k (or the array of these elements) are called the *components* of z with respect to the basis $\mathscr{E}$. (Precisely speaking, they should be called the components with respect to the standard basis corresponding to $\mathscr{E}$.)

When we write the components, the upper indices correspond to V and the lower ones to V^*. Thus, in the case of $\xi_{j_1, \ldots, j_q}^{\quad i_1, \ldots, i_p}$, the $i_1, \ldots, i_p$

correspond to V and the $j_1, \ldots, j_q$ to V^*. Usually indices are written one over another like $\xi^{i,\ldots}_{j,\ldots}$, but in order to read them easily, we shift them as above. When we specify the order of V and V^*, for instance to express that the components are of a tensor of type $(1, -1, 1)$, we denote them by $\xi^{i_1}{}_{j_1}{}^{i_2}$ (or $\xi^{i_1 \cdot i_2}_{\cdot j_1 \cdot}$).

EXAMPLE 2.5. Linear transformations of V can be considered as tensors of type $(1, 1)$ (§1, Example 2.1). Let us compute the components of $T \in \operatorname{Hom}(V, V)$ with respect to $\mathscr{E}$. Let (α_{ij}) be the matrix associated with T with respect to the basis $\mathscr{E} = (e_1, \ldots, e_n)$. That is, $T(e_k) = \sum_i \alpha_{ik} e_i$.

Let $(t_i{}^j)$ be the standard basis of $T_1{}^1(V) = V \otimes V^*$ corresponding to $\mathscr{E}$. Then tensor t_T corresponding to T is expressed as

$$t_T = \sum_{i,j} \xi^i{}_j t_i{}^j,$$

where $\{\xi^i{}_j\}$ are the components of t_T. The linear transformation T_{f^j, e_i} corresponding to the basis element $t_i{}^j = e_i \otimes f^j$ is given by

$$T_{f^j, e_i}(v) = f^j(v) e_i \qquad (v \in V).$$

Therefore

$$\begin{aligned} T(e_k) &= \sum_{i,j} \xi^i{}_j T_{f^j, e_i}(e_k) = \sum_{i,j} \xi^i{}_j f^j(e_k) e_i \\ &= \sum_{i,j} \xi^i{}_j \delta_{jk} e_i = \sum_{i,j} \xi^i{}_k e_i. \end{aligned}$$

Thus we have $\xi^i{}_k = \alpha_{ik}$, that is, the component $\xi^i{}_j$ is the component α_{ij} of the matrix.

EXAMPLE 2.6. Bilinear forms Φ on V can be considered as covariant tensors of degree 2 (§1, Example 2.2). With respect to the standard basis of $T_2{}^0(V)$ $\Phi = \sum_i \sum_j \xi_{ij} t^{ij}$. (To simplify the notation, we identity $\mathscr{L}(V, V; k)$ and $V^* \otimes V^*$.) Then

$$\begin{aligned} \Phi(e_k, e_l) &= \sum_i \sum_j \xi_{ij} t^{ij}(e_k, e_l) = \sum_i \sum_j \xi_{ij} f^i(e_k) f^j(e_l) \\ &= \sum_i \sum_j \xi_{ij} \delta_{ik} \delta_{jl} = \xi_{kl}. \end{aligned}$$

Thus the component ξ_{ij} is given by $\Phi(e_i, e_j)$.

In general, the components $\{\xi_{j_1, \ldots, j_q}\}$ of a q-multilinear form Φ on V with respect to $\mathscr{E}$ are given by

$$\xi_{j_1, \ldots, j_q} = \Phi(e_{j_1}, \ldots, e_{j_q}).$$

Similarly, the components of an element

$$\Phi \in \mathscr{L}(\underbrace{V, \ldots, V}_{q \text{ factors}}, \underbrace{V^*, \ldots, V^*}_{p \text{ factors}}; k)$$

with respect to $\mathscr{E}$ are given by

$$\xi_{j_1,\dots,j_q}{}^{i_1,\dots,i_p} = \Phi(e_{j_1}, \dots, e_{j_q}, f^{i_1}, \dots, f^{i_p}).$$

EXERCISE. Prove these formulas.

Next, we study the relation between the components when we take another basis $\tilde{\mathscr{E}} = (\tilde{e}_1, \dots, \tilde{e}_n)$ for V. Let $F = (\alpha^i{}_j)$ be the matrix of the change of bases from $\mathscr{E}$ to $\tilde{\mathscr{E}}$. Namely, we have

$$\tilde{e}_j = \sum_i \alpha^i{}_j e_i \qquad (j = 1, \dots, n). \tag{2.1}$$

First we compute the matrix of the change of dual bases. Let $(\tilde{f}^1, \dots, \tilde{f}^n)$ be the dual basis for $\tilde{\mathscr{E}}$. Since we have

$$f^i(\tilde{e}_j) = f^i\left(\sum_l \alpha^l{}_j e_l\right) = \sum_l \alpha^l{}_j f^i(e_l) = \alpha^i{}_j,$$

$$f^i = \sum_j \alpha^i{}_j \tilde{f}^j \qquad (i = 1, \dots, n). \tag{2.2}$$

Let $F^{-1} = (\beta^i{}_j)$. By (2.1)

$$e_i = \sum_i \beta^i{}_j \tilde{e}_i \qquad (j = 1, \dots, n). \tag{2.3}$$

Let $(\tilde{t}_{i_1,\dots,i_p}{}^{j_1,\dots,j_q})$ be the standard basis of $T_q{}^p(V)$ corresponding to the basis $\tilde{\mathscr{E}}$ of V and $\{\tilde{\xi}_{j_1,\dots,j_q}{}^{i_1,\dots,i_p}\}$ the components of $z \in T_q{}^p(V)$ with respect to $\tilde{\mathscr{E}}$.

PROPOSITION 2.1. *With the notation as above, we have the following formulas relating the components* $\{\xi_{j_1,\dots,j_q}{}^{i_1,\dots,i_p}\}$ *and* $\{\tilde{\xi}_{j_1,\dots,j_q}{}^{i_1,\dots,i_p}\}$. *For all* i_ν *and* j_μ $(1 \le i_\nu \le n,\ 1 \le j_\mu \le n)$,

$$\tilde{\xi}_{j_1,\dots,j_q}{}^{i_1,\dots,i_p} = \sum_{k_1,\dots,k_p,l_1,\dots,l_q} \alpha^{l_1}{}_{j_1} \cdots \alpha^{l_q}{}_{j_q} \beta^{i_1}{}_{k_1} \cdots \beta^{i_p}{}_{k_p} \xi_{l_1,\dots,l_q}{}^{k_1,\dots,k_p}. \tag{2.4}$$

PROOF. By definition, we have

$$\begin{aligned} z &= \sum_{k_1,\dots,k_p,l_1,\dots,l_q} \xi_{l_1,\dots,l_q}{}^{k_1,\dots,k_p} t_{k_1,\dots,k_p}{}^{l_1,\dots,l_q} \\ &= \sum_{k_1,\dots,k_p,l_1,\dots,l_q} \xi_{l_1,\dots,l_q}{}^{k_1,\dots,k_p} e_{k_1} \otimes \cdots \otimes e_{k_p} \otimes f^{l_1} \otimes \cdots \otimes f^{l_q}. \end{aligned}$$

(Substitute (2.2) and (2.3).)

$$
\begin{aligned}
z &= \sum_{k_1,\dots,k_p,l_1,\dots,l_q} \xi_{l_1,\dots,l_q}{}^{k_1,\dots,k_p} \left(\sum_{i_1} \beta^{i_1}{}_{k_1}\tilde{e}_{i_1}\right) \otimes \cdots \otimes \left(\sum_{j_q} \alpha^{l_q}{}_{j_q}\tilde{f}^{j_q}\right) \\
&= \sum_{i_1,\dots,i_p,j_1,\dots,j_q,} \ \sum_{k_1,\dots,k_p,l_1,\dots,l_q} \beta^{i_1}{}_{k_1} \cdots \beta^{i_p}{}_{k_p} \\
&\quad \times \alpha^{l_1}{}_{j_1} \cdots \alpha^{l_q}{}_{j_q} \xi_{l_1,\dots,l_q}{}^{k_1,\dots,k_p} \tilde{e}_{i_1} \otimes \cdots \otimes \tilde{e}_{i_p} \otimes \tilde{f}^{j_1} \otimes \cdots \otimes \tilde{f}^{j_q} \\
&= \sum_{i_1,\dots,i_p,j_1,\dots,j_q} \left(\sum_{k_1,\dots,k_p,l_1,\dots,l_q} \alpha^{l_1}{}_{j_1} \cdots \alpha^{l_q}{}_{j_q} \beta^{i_1}{}_{k_1} \cdots \beta^{i_p}{}_{k_p} \right. \\
&\qquad\qquad \left. \times \xi_{l_1,\dots,l_q}{}^{k_1,\dots,k_p} \right) \tilde{t}_{i_1,\dots,i_p}{}^{j_1,\dots,j_q}.
\end{aligned}
$$

On the other hand,

$$z = \sum \tilde{\xi}_{j_1,\dots,j_q}{}^{i_1,\dots,i_p} \tilde{t}_{i_1,\dots,i_p}{}^{j_1,\dots,j_q}.$$

Making a comparison between the coefficients of $\tilde{t}_{i_1,\dots,i_p}{}^{j_1,\dots,j_q}$, we obtain the formula. □

Formula (2.4) of Proposition 2.1 is called the *transformation law of tensor components*. In particular, the transformation law of the components of contravariant vectors and covariant vectors are

$$\tilde{\xi}^i = \sum_k \beta^i{}_k \xi^k, \qquad \tilde{\xi}_j = \sum_l \alpha^l{}_j \xi_l.$$

The next proposition is, in a sense, the converse of Proposition 2.1.

PROPOSITION 2.2. *To every basis of* V *there corresponds an array* $\{\xi_{j_1,\dots,j_q}{}^{i_1,\dots,i_p} | 1 \le i_\nu \le n,\ 1 \le j_\mu \le n\}$ *of* n^{p+q} *elements of* k *such that, if the matrix for the change of bases from* $\mathscr{E}$ *to* $\tilde{\mathscr{E}}$ *is given by* $(\alpha^i{}_j)$ *(i.e.,* $\tilde{\xi}_j = \sum_i \alpha^i{}_j e_i$*), we have the same formulas as in (2.4) relating the elements of the arrays corresponding to* $\mathscr{E}$ *and* $\tilde{\mathscr{E}}$*, where* $(\beta^i{}_j) = (\alpha^i{}_j)^{-1}$*. Then, there exists a tensor* z *of type* (p, q) *such that the components of* z *with respect to* $\mathscr{E}$ *are given by* $\{\xi_{j_1,\dots,j_q}{}^{i_1,\dots,i_p}\}$.

PROOF. Let $\mathscr{E} = (e_i)$ be a basis of V. If the array $\{\xi_{j_1,\dots,j_q}{}^{i_1,\dots,i_p}\}$ corresponds to the basis $\mathscr{E}$, define $z \in T_q{}^p(V)$ by

$$z = \sum \xi_{j_1,\dots,j_q}{}^{i_1,\dots,i_p} t_{i_1,\dots,i_p}{}^{j_1,\dots,j_q},$$

where $(t_{i_1,\dots,i_p}{}^{j_1,\dots,j_q})$ is the standard basis corresponding to $\mathscr{E}$. By Propostion 2.1, we know that the components of z with respect to any basis coincide with the array given previously. □

This proposition has special interest, since it relates to the classical definition of tensors. Classically, a tensor was defined as an object described in the assumption of Proposition 2.2, that is, the family of arrays satisfying (2.4), corresponding to bases of V. This proposition shows that tensors defined classically are nothing but the components of tensors. In many cases, tensors which appear in applications can be recognized by the following procedure: we have an object whose components with respect to each coordinate system (basis) are given. Then we check the transformation law of the components and we know whether this object is a tensor or not. We give a simple example.

EXAMPLE 2.7. To every basis, there corresponds n^2 elements of $\{\delta'^i_{\ j} | 1 \le i \le n,\ 1 \le j \le n\}$ (where $\delta'^i_{\ j} = \delta_{ij}$ is Kronecker's symbol). Then these are the components of a tensor of type $(1, 1)$.

In fact, let $\mathscr{E} = (e_i)$ and $\widetilde{\mathscr{E}} = (\tilde{e}_i)$ be bases such that $\tilde{e}_j = \sum \alpha'^i_{\ j} e_i$. Then we can verify (2.4) as follows.

$$\sum_{k,l} \alpha'^l_{\ j} \beta'^i_{\ k} \delta'^k_{\ l} = \sum_l \alpha'^l_{\ j} \beta'^i_{\ l} = \delta'^i_{\ j},$$

(where $(\beta'^i_{\ j}) = (\alpha'^i_{\ j})^{-1}$).

A tensor of type $(1, 1)$ can be identified with a linear transformation of V. The tensor above is identified with the identity mapping. The phenomenon above reflects the fact that the matrix of the identity mapping with respect to any basis is the unit matrix.

(b) Products and contractions of tensors. Since the tensor space $T_q^{\ p}(V)$ is a k-vector space, addition and scalar multiplication are defined on $T_q^{\ p}(V)$. Moreover we can define products of tensors of different types. This operation plays a fundamental role in the definition of a tensor algebra (§4). First we describe this product.

Starting with the k-vector spaces $T_q^{\ p}(V)$ and $T_s^{\ r}(V)$, we can construct a tensor product $T_q^{\ p}(V) \otimes T_s^{\ r}(V)$, which can be naturally identified with $T_{q+s}^{\ \ p+r}(V)$ (see §1). Thus the bilinear mapping ι which defines the tensor product

$$\begin{array}{ccc} T_q^{\ p}(V) \times T_s^{\ r}(V) & \to & T_q^{\ p}(V) \otimes T_s^{\ r}(V) \\ \Psi & & \Psi \\ (x, y) & \mapsto & x \otimes y \end{array}$$

can be considered as bilinear mapping into $T_{q+s}^{\ \ p+r}(V)$. The image of $(x, y) \in T_q^{\ p}(V) \times T_s^{\ r}(V)$ under this mapping is called the *product* of x and y and is denoted by xy or $x \otimes y$.

Let $\mathscr{E}$ be a basis of V. Take the standard bases corresponding to $\mathscr{E}$ of $T_q^{\ p}(V)$, $T_s^{\ r}(V)$, and $T_{q+s}^{\ \ p+r}(V)$. Let $\{\xi_{j_1, \dots, j_q}^{\ \ i_1, \dots, i_p}\}$ and $\{\eta_{l_1, \dots, l_s}^{\ \ k_1, \dots, k_r}\}$ be the components of x and y with respect to these bases respectively. It is easily verified that the components of xy are given by $\{\xi_{j_1, \dots, j_q}^{\ \ i_1, \dots, i_p} \times \eta_{l_1, \dots, l_s}^{\ \ k_1, \dots, k_r}\}$.

Next, we shall define the contraction of a tensor. We use the following proposition.

PROPOSITION 2.3. *Consider the tensor space* $T_q^{\,p}(V)$ *and assume that* $p > 0$ *and* $q > 0$. *Fix integers* r *and* s *such that* $1 \le r \le p$ *and* $1 \le s \le q$. *Then there is a unique linear mapping*

$$C_s^{\,r}: T_q^{\,p}(V) \to T_{q-1}^{\,p-1}(V),$$

such that, for all $v_i \in V$ *and* $\varphi_j \in V^*$, *we have* (1)

$$C_s^{\,r}(v_1 \otimes \cdots \otimes v_p \otimes \varphi_1 \otimes \cdots \otimes \varphi_q) \\ = \varphi_s(v_r) v_1 \otimes \cdots \otimes \hat{v}_r \otimes \cdots \otimes v_p \otimes \varphi_1 \otimes \cdots \otimes \hat{\varphi}_s \otimes \cdots \otimes \varphi_q .$$

PROOF. It is easy to see that the mapping

$$\begin{array}{ccc} V \times \cdots \times V \times V^* \times \cdots \times V^* & \longrightarrow & T_{q-1}^{\,p-1}(V) \\ \Psi & & \Psi \\ (v_1, \ldots, v_p, \varphi_1, \ldots, \varphi_q) & \mapsto & \varphi_s(v_r) v_1 \otimes \cdots \otimes \hat{v}_r \otimes \cdots\cdots \otimes \hat{\varphi}_s \otimes \cdots \otimes \varphi_q \end{array}$$

is a $(p+q)$-multilinear mapping. Thus, by the fundamental property (T) of tensor products, this mapping induces a unique linear mapping $C_s^{\,r}: T_q^{\,p}(V) \to T_{q-1}^{\,p-1}(V)$ with required property. □

DEFINITION 2.1. The linear mapping $C_s^{\,r}$, the existence of which is proved by Proposition 2.3, is called the *contraction* with respect to rth contravariant index and sth covariant index.

If the components of $z \in T_q^{\,p}(V)$ with respect to a basis $\mathscr{E}$ of V are given by $\{\xi_{j_1,\ldots,j_q}{}^{i_1,\ldots,i_p}\}$, the components $\{\eta_{j_1,\ldots,j_{q-1}}{}^{i_1,\ldots,i_{p-1}}\}$ of $C_s^{\,r}(z)$ are given by

$$\eta_{j_1,\ldots,j_{q-1}}{}^{i_1,\ldots,i_{p-1}} = \sum_{\rho=1}^{n} \xi_{j_1,\ldots,j_{s-1},\rho,j_s,\ldots,j_{q-1}}{}^{i_1,\ldots,i_{r-1},\rho,i_r,\ldots,i_{p-1}}.$$

In fact, this follows from formula (2.5) for the standard basis element $t_{i_1,\ldots,i_p}{}^{j_1,\ldots,j_q}$.

$$C_s^{\,r}(t_{i_1,\ldots,i_p}{}^{j_1,\ldots,j_q}) = \begin{cases} t_{i_1,\ldots,\hat{i}_r,\ldots,i_p}{}^{j_1,\ldots,\hat{j}_s,\ldots,j_q} & (\text{if } i_r = j_s), \\ 0 & (\text{if } i_r \ne j_s). \end{cases} \tag{2.5}$$

REMARK. When we make a summation from 1 to n $(= \dim V)$ with respect to an upper index and a lower index as in (2.4) or in the formula

$$\sum \xi_{j_1,\ldots,j_{s-1},\rho,j_s,\ldots,j_{q-1}}{}^{i_1,\ldots,i_{r-1},\rho,i_r,\ldots,i_{p-1}}$$

above, some authors omit the summation sign Σ. For example, (2.4) is written as

$$\tilde{\xi}_{j_1,\ldots,j_q}{}^{i_1,\ldots,i_p} = \alpha^{l_1}{}_{j_1} \cdots \alpha^{l_q}{}_{j_q} \beta^{i_1}{}_{k_1} \cdots \beta^{i_p}{}_{k_p} \xi_{l_1,\ldots,l_q}{}^{k_1,\ldots,k_p}.$$

(1) By $\hat{v}_r$ and $\hat{\varphi}_s$, we indicate that these factors are missing.

This convention is often used in tensor calculus and is called *Einstein's convention* after the originator. This notation is convenient, but we do not use it in this book in order to avoid any misunderstanding.

EXAMPLE 2.8. By the isomorphism $\mathrm{Hom}(V, V) \cong T_1{}^1(V) = V \otimes V^*$, let $T \in \mathrm{Hom}(V, V)$ correspond to $t_T \in V \otimes V^*$. Then, for $t_T \in V \otimes V^*$,

(1) $C_2{}^1(t_T) = \mathrm{tr}(T)$, where $C_2{}^1$ means the contraction with respect to the first contravariant index and the next covariant index. In this example, we use the same convention.

(2) For any $v \in V$, $T(v) = C_2{}^3(t_T \otimes v)$, where $t_T \otimes v \in V \otimes V^* \otimes V$.

(3) Let $T, S \in \mathrm{Hom}(V, V)$ and $R = T \circ S$. Then, $t_R = C_2{}^3(t_T \otimes t_S)$, where $t_T \otimes t_S \in V \otimes V^* \otimes V \otimes V^*$.

The equality in (1) is obtained from the formula for the components of t_T (§2, Example 2.5). Set $t_T = \sum v_i \otimes \varphi_i$. Then (2) follows from it by definition. Similarly we obtain (3).

EXAMPLE 2.9 (Quotient rule of tensor). Assume that, to every basis $\mathscr{E}$ of V, there corresponds n^{p+q} elements $\{\xi_{j_1, \dots, j_q}{}^{i_1, \dots, i_p} \mid 1 \le i_\nu \le n,\ 1 \le j_\mu \le n\}$ of k. Moreover, assume that there exist s and t $(1 \le s \le q,\ q \le t \le p)$, for which the following condition is satisfied; for any tensor y of type (s, t), there exists a tensor z_y of type $(p-t, q-s)$ such that the components of z_y with respect to $\mathscr{E}$ are given by

$$\zeta_{j_{s+1}, \dots, j_q}{}^{i_{t+1}, \dots, i_p} = \sum_{l_1, \dots, l_t, k_1, \dots, k_s} \xi_{k_1, \dots, k_s, j_{s+1}, \dots, j_q}{}^{l_1, \dots, l_t, i_{t+1}, \dots, i_p} \eta_{l_1, \dots, l_t}{}^{k_1, \dots, k_s}$$

where the $\{\eta_{l_1, \dots, l_t}{}^{k_1, \dots, k_s}\}$ are the components of y with respect to $\mathscr{E}$. Then there exists a tensor x whose components with respect to $\mathscr{E}$ are given by $\{\xi_{j_1, \dots, j_q}{}^{i_1, \dots, i_p}\}$.

PROOF. Let $(t_{i_1, \dots, i_p}{}^{j_1, \dots, j_p})$ be the standard basis of $T_q{}^p(V)$ corresponding to $\mathscr{E}$. Define $x \in T_q{}^p(V)$ by

$$x = \sum_{i_1, \dots, i_p, j_1, \dots, j_q} \xi_{j_1, \dots, j_q}{}^{i_1, \dots, i_p} t_{i_1, \dots, i_p}{}^{j_1, \dots, j_q}.$$

Then, comparing the components with respect to $\mathscr{E}$, we know that, for all $y \in T_t^s(V)$, (²)

$$z_y = c_{q+1}{}^1 c_{q+2}{}^2 \cdots c_{q+t}{}^t c_1{}^{p+1} \cdots c_s^{p+s}(x \otimes y).$$

Let $\tilde{\mathscr{E}}$ be a basis of V and $\{\tilde{\xi}_{j_1, \dots, j_q}{}^{i_1, \dots, i_p}\}$ be the given elements of k corresponding to $\tilde{\mathscr{E}}$. Let $\{\tilde{\xi}'{}_{j_1, \dots, j_q}{}^{i_1, \dots, i_p}\}$ and $\{\tilde{\eta}_{l_1, \dots, l_t}{}^{k_1, \dots, k_s}\}$ be the components of x and y respectively with respect to $\tilde{\mathscr{E}}$. Then, comparing the

(²) Here, the indices of contraction are the original indices in $x \otimes y \in T_{q+t}{}^{p+s}(V)$.

components of z_y we have

$$\sum_{l_1,\dots,l_t,k_1,\dots,k_s} \tilde{\xi}_{k_1,\dots,k_s,j_{s+1},\dots,j_q}{}^{l_1,\dots,l_t,i_{t+1},\dots,i_p}\tilde{\eta}_{l_1,\dots,l_t}{}^{k_1,\dots,k_s}$$
$$= \sum_{l_1,\dots,l_t,k_1,\dots,k_s} \tilde{\xi}'{}_{k_1,\dots,k_s,j_{s+1},\dots,j_q}{}^{l_1,\dots,l_t,i_{t+1},\dots,i_p}\tilde{\eta}_{l_1,\dots,l_t}{}^{k_1,\dots,k_s}.$$

Since this is true for all y, we can take as $y = \tilde{t}_{j_1,\dots,j_s}{}^{i_1,\dots,i_t}$ (standard basis element corresponding to $\tilde{\mathscr{E}}$) and we obtain

$$\tilde{\xi}_{j_1,\dots,j_q}{}^{i_1,\dots,i_p} = \tilde{\xi}'{}_{j_1,\dots,j_q}{}^{i_1,\dots,i_p},$$

which shows that x satisfies the condition of the assertion. □

As an easy example of quotient rule, let us examine a linear transformation, though we already know that it is a tensor. Let $T \in \mathrm{Hom}(V, V)$. To every basis of V, there corresponds the elements $(\alpha^i{}_j)$ of the matrix for T with respect to that basis. For any (contravariant) vector v (= tensor of type $(1, 0)$), consider the vector $w = T(v)$. If the $\{v^i\}$ are the components of v, the components of w are given by $\sum_l \alpha^i{}_l v^l$, which satisfies the assumption of the quotient rule, for $(s, t) = (1, 0)$ and $(p-t, q-s) = (1, 0)$. By the quotient rule the $\{\alpha^i{}_j\}$ are the components of a tensor of type $(1, 1)$.

§3. Symmetric tensors and alternating tensors

(a) Definitions. There are families of tensors which are called symmetric or alternating. In this section, we give their definitions and study their properties. First, we consider the space $T^p(V)$ of contravariant tensors of degree p. In this section, we write $T^p(V) = T^p$ for simplicity.

Let $\mathfrak{S}_p$ be the set of permutations of the set $\{1, \dots, p\}$ with p elements. Denote by $\mathrm{sgn}(\sigma)$ the signature of $\sigma \in \mathfrak{S}_p$ (i.e., $\mathrm{sgn}(\sigma) = 1$ if σ is an even permutation and $\mathrm{sgn}(\sigma) = -1$ if σ is an odd permutation.)

From the next proposition we know that there corresponds to every element σ of $\mathfrak{S}_p$ a linear transformation P_σ of T^p.

Proposition 2.4. (1) *Let $\sigma \in \mathfrak{S}_p$. There exists a unique linear transformation P_σ of T^p such that*

$$P_\sigma(v_1 \otimes \cdots \otimes v_p) = v_{\sigma^{-1}(1)} \otimes \cdots \otimes v_{\sigma^{-1}(p)} \qquad (v_i \in V). \tag{2.6}$$

Moreover, P_σ is a linear isomorphism.

(2) *For $\sigma, \tau \in \mathfrak{S}_p$, $P_\sigma P_\tau = P_{\sigma\tau}$.*

Denote by 1 *the identity permutation. Then*

$$P_1 = I \qquad (= \text{the identity transformation of } T^p).$$

Proof. (1) To prove the existence and the uniqueness of P_σ, consider a

p-multilinear mapping defined as follows;

$$\begin{array}{ccc} V \times \cdots \times V & \longrightarrow & T^p \\ \Psi & & \Psi \\ (v_1, \ldots, v_p) & \mapsto & v_{\sigma^{-1}(1)} \otimes \cdots \otimes v_{\sigma^{-1}(p)}. \end{array}$$

From property (T) of tensor products, this mapping induces a linear mapping $P_\sigma : T^p \to T^p$. Since P_σ induces a permutation of the standard basis elements $(t_{i_1, \ldots, i_p})$, it is a linear isomorphism.

(2) We have

$$\begin{aligned} P_\sigma P_\tau(v_1 \otimes \cdots \otimes v_p) &= P_\sigma(v_{\tau^{-1}(1)} \otimes \cdots \otimes v_{\tau^{-1}(p)}) \\ &= v_{\tau^{-1}(\sigma^{-1}(1))} \otimes \cdots \otimes v_{\tau^{-1}(\sigma^{-1}(p))} \\ &= P_{\sigma\tau}(v_1 \otimes \cdots \otimes v_p). \end{aligned}$$

Since all the elements of $\{v_1 \otimes \cdots \otimes v_p | v_i \in V\}$ generate T^p, we know that $P_\sigma P_\tau = P_{\sigma\tau}$. Clearly $P_1 = I$. □

REMARK. We can prove existence as follows: define P_σ on the standard basis elements by

$$P_\sigma(t_{i_1, \ldots, i_p}) = t_{j_1, \ldots, j_p} \quad (\text{where } j_k = i_{\sigma^{-1}(k)}),$$

then extend P_σ to the whole of T^p by linearity. In this case, we must verify (2.6) and show that the definition of P_σ does not depend on the choice of bases.

EXERCISE 1. Let $t \in T^p(V)$ and $\{\xi^{i_1, \ldots, i_p}\}$ be the components of t with respect to $\mathscr{E}$. Show that $\{\xi^{i_{\sigma(1)}, \ldots, i_{\sigma(p)}}\}$ are the components of $P_\sigma(t)$.

DEFINITION 2.2. Let $t \in T^p$. If $P_\sigma(t) = t$ for all $\sigma \in \mathfrak{S}_p$, t is called a *symmetric tensor*. If $P_\sigma(t) = \operatorname{sgn}(\sigma)t$ for all $\sigma \in \mathfrak{S}_p$, then t is called an *alternating* (or skew-symmetric) *tensor*.

The set of symmetric tensors and that of alternating tensors are vector subspaces of T^p and are denoted by $S^p(V)$ (or simply S^p) and $A^p(V)$ (or simply A^p) respectively.

EXAMPLE 2.10. If $p = 1$, $S^1(V) = A^1(V) = T^1 = V$.

If $p = 2$, since $\mathfrak{S}_2 = \{1, (1\ 2)\}$,

$$\begin{aligned} t \in S^2(V) &\quad \Leftrightarrow \quad P_{(1\ 2)}(t) = t, \\ t \in A^2(V) &\quad \Leftrightarrow \quad P_{(1\ 2)}(t) = -t. \end{aligned}$$

Thus, $T^2(V) = S^2(V) \oplus A^2(V)$ (direct sum). For a standard basis element t_{ij}, we have $P_{(1\ 2)}(t_{ij}) = t_{ji}$. Therefore we know that $\{t_{ij} + t_{ji}\ (i \le j)\}$ is a basis for $S^2(V)$ and $\{t_{ij} - t_{ji}\ \ (i < j)\}$ is a basis for $A^2(V)$. In particular, if $\dim V = n$,

$$\dim S^2(V) = n(n+1)/2, \qquad \dim A^2(V) = n(n-1)/2.$$

EXAMPLE 2.11. From Exercise 1 and the definition, we obtain the following assertions. We denote by $\{\xi^{i_1,\dots,i_p}\}$ the components of $t \in T^p(V)$ with respect to a basis for V.

(1) t is a symmetric tensor

$$\Leftrightarrow \quad \xi^{i_{\sigma(1)},\dots,i_{\sigma(P)}} = \xi^{i_1,\dots,i_p} \quad \text{for all } \sigma \in \mathfrak{S}_p.$$

$$\Leftrightarrow \quad \xi^{i_{\sigma(1)},\dots,i_{\sigma(P)}} = \xi^{i_1,\dots,i_p} \quad \text{for all transpositions } \sigma \in \mathfrak{S}_p.$$

(2) t is an alternating tensor

$$\Leftrightarrow \quad \xi^{i_{\sigma(1)},\dots,i_{\sigma(P)}} = \operatorname{sgn}(\sigma)\xi^{i_1,\dots,i_p} \quad \text{for all } \sigma \in \mathfrak{S}_p.$$

$$\Leftrightarrow \quad \xi^{i_{\sigma(1)},\dots,i_{\sigma(P)}} = -\xi^{i_1,\dots,i_p} \quad \text{for all transpositions } \sigma \in \mathfrak{S}_p.$$

EXERCISE 2. Verify these assertions.

Let us examine the properties of subspaces $S^p(V)$ and $A^p(V)$. For this purpose, define the linear transformations $\mathscr{S}_p$ and $\mathscr{A}_p$ by the following formulas.

$$\mathscr{S}_p = \frac{1}{p!}\sum_{\sigma\in\mathfrak{S}_p} P_\sigma, \qquad \mathscr{A}_p = \frac{1}{p!}\sum_{\sigma\in\mathfrak{S}_p} \operatorname{sgn}(\sigma)P_\sigma.$$

When there is no risk of confusion, we simply write $\mathscr{S}_p = \mathscr{S}$ and $\mathscr{A}_p = \mathscr{A}$, omitting p. The next lemma follows easily from the definition.

LEMMA 2.1. *For $\sigma \in \mathfrak{S}_p$,*

$$P_\sigma\mathscr{S} = \mathscr{S}P_\sigma = \mathscr{S}, \qquad P_\sigma\mathscr{A} = \mathscr{A}P_\sigma = \operatorname{sgn}(\sigma)\mathscr{A}.$$

PROOF. $P_\sigma\mathscr{S} = (1/p!)P_\sigma(\sum_{\tau\in\mathfrak{S}_p} P_\tau) = (1/p!)\sum_{\tau\in\mathfrak{S}_p} P_\sigma P_\tau = (1/p!)\sum_{\tau\in\mathfrak{S}_p} P_{\sigma\tau}$. Now, if τ runs over $\mathfrak{S}_p$, the same holds for $\sigma\tau$. Thus,

$$P_\sigma\mathscr{S} = \frac{1}{p!}\sum_{\tau\in\mathfrak{S}_p} P_\tau = \mathscr{S}.$$

Similarly,

$$P_\sigma\mathscr{A} = \frac{1}{p!}P_\sigma\left(\sum_{\tau\in\mathfrak{S}_p}\operatorname{sgn}(\tau)P_\tau\right) = \frac{1}{p!}\left(\sum_{\tau\in\mathfrak{S}_p}\operatorname{sgn}(\tau)P_\sigma P_\tau\right).$$

Since $\operatorname{sgn}(\sigma\tau) = \operatorname{sgn}(\sigma)\operatorname{sgn}(\tau)$ and $\operatorname{sgn}(\sigma) = \pm 1$, notice that $\operatorname{sgn}(\tau) = \operatorname{sgn}(\sigma)\operatorname{sgn}(\sigma\tau)$, and $\sigma\tau$ runs over $\mathfrak{S}_p$. Thus

$$P_\sigma\mathscr{A} = \operatorname{sgn}(\sigma)\mathscr{A}.$$

The other equalities can be obtained by similar reasoning. □

Applying this lemma, we have the next proposition.

PROPOSITION 2.5. *The linear mappings $\mathscr{S}$ and $\mathscr{A}$ are projections of T^p onto the vector subspaces S^p and A^p respectively. Namely,*

(1) $\mathscr{S}^2 = \mathscr{S}$, $\mathscr{A}^2 = \mathscr{A}$.

(2) *For* $t \in T^p$,

$$t \in S^p \Leftrightarrow \mathscr{S}(t) = t, \qquad t \in A^p \Leftrightarrow \mathscr{A}(t) = t.$$

Moreover, if $p \geq 2$, $\mathscr{S}\mathscr{A} = \mathscr{A}\mathscr{S} = 0$, *which implies* $S^p \cap A^p = \{0\}$.

NOTE. Let X be a vector space and Y a subspace of X. Let Y' be a subspace of X such that $X = Y \oplus Y'$ (direct sum). For $x \in X$, decompose it as $x = y + y'$ $(y \in Y,\ y' \in Y')$. The linear transformation P of X defined by $P(x) = y$ is called a projection of X onto Y (defined by Y'). In general, a linear transformation defined by some Y' is called a *projection* onto Y. A linear transformation is a projection onto Y if and only if the following hold.

(1) $P^2 = P$.

(2) If $x \in X$, then $x \in Y$ if and only if $P(x) = x$.

PROOF. (1)

$$\mathscr{S}^2 = \left(\frac{1}{p!}\sum_{\sigma \in \mathfrak{S}_p} P_\sigma\right)\mathscr{S} = \frac{1}{p!}\sum_{\sigma \in \mathfrak{S}_p} \mathscr{S} = \mathscr{S},$$

$$\mathscr{A}^2 = \left(\frac{1}{p!}\sum_{\sigma \in \mathfrak{S}_p} \operatorname{sgn}(\sigma) P_\sigma\right)\mathscr{A} = \frac{1}{p!}\sum_{\sigma \in \mathfrak{S}_p} (\operatorname{sgn}(\sigma))^2 \mathscr{A} = \mathscr{A}.$$

(2) By definition,

$$t \in S^p \Leftrightarrow P_\sigma(t) = t \quad \text{for all } \sigma \in \mathfrak{S}_p.$$

Thus, if $t \in S^p$, $\mathscr{S}(t) = (1/p!)\sum P_\sigma(t) = t$. Conversely, if $\mathscr{S}(t) = t$, $P_\sigma(t) = P_\sigma \mathscr{S}(t) = \mathscr{S}(t) = t$ for all P_σ, and $t \in S^p$.

$$t \in A^p \Leftrightarrow P_\sigma(t) = \operatorname{sgn}(\sigma) t \quad \text{for all } \sigma \in \mathfrak{S}_p.$$

Thus, if $t \in A^p$, $\mathscr{A}(t) = (1/p!)\Sigma \operatorname{sgn}(\sigma) P_\sigma(t) = (1/p!)\Sigma(\operatorname{sgn}(\sigma))^2 t = t$. Conversely, if $\mathscr{A}(t) = t$, $P_\sigma(t) = P_\sigma \mathscr{A}(t) = \operatorname{sgn}(\sigma)\mathscr{A}(t) = \operatorname{sgn}(\sigma) t$ for all P_σ, and $t \in A^p$.

Moreover,

$$\mathscr{S}\mathscr{A} = \left(\frac{1}{p!}\sum_{\sigma \in \mathfrak{S}_p} P_\sigma\right)\mathscr{A} = \frac{1}{p!}\sum_{\sigma \in \mathfrak{S}_p} \operatorname{sgn}(\sigma)\mathscr{A} = \frac{1}{p!}\left(\sum_{\sigma \in \mathfrak{S}_p} \operatorname{sgn}(\sigma)\right)\mathscr{A}.$$

Since the number of even permutations is equal to the number of odd permutations if $p \geq 2$, we have $\sum_{\sigma \in \mathfrak{S}_p} \operatorname{sgn}(\sigma) = 0$ and $\mathscr{S}\mathscr{A} = 0$. Similarly, $\mathscr{A}\mathscr{S} = 0$.

Suppose that $t \in S^p \cap A^p$. $t \in S^p$ implies $t = S(t)$. Applying $\mathscr{A}$ on both sides, $\mathscr{A}(t) = \mathscr{A}\mathscr{S}(t) = 0$. On the other hand, $t \in A^p$ implies $t \in \mathscr{A}(t)$. So $t = 0$. □

This proposition shows that $S^p = \operatorname{Im}\mathscr{S}$ and $A^p = \operatorname{Im}\mathscr{A}$. In particular, for any $t \in T^p$, $\mathscr{S}(t)$ is a symmetric tensor and $\mathscr{A}(t)$ is an alternating tensor. Therefore, the mappings $\mathscr{S}$ and $\mathscr{A}$ are called the *symmetrizer* and the *alternator* on T^p.

(b) Bases and dimensions of the symmetric tensor space and the alternating tensor space. Starting from a standard basis $(t_{i_1, \dots, i_p})$ $(1 \leq i_\nu \leq n)$ of T (corresponding to a basis $\mathscr{E}$ of V) and applying the symmetrizer and the alternator, we are going to construct bases of S^p and A^p. To read indices more easily, we write $t_{i_1, \dots, i_p} = t(i_1, \dots, i_p)$. By definition, it follows that, for $\sigma \in \mathfrak{S}_p$,

$$P_\sigma(t(i_1, \dots, i_p)) = t(i_{\sigma^{-1}(1)}, \dots, i_{\sigma^{-1}(p)}).$$

From previous observation, we know that the spaces S^p and A^p are generated by

$$\{\mathscr{S}(t(i_1, \dots, i_p)) | 1 \leq i_\nu \leq n\} \quad \text{and} \quad \{\mathscr{A}(t(i_1, \dots, i_p)) | 1 \leq i_\nu \leq n\}$$

respectively. Thus we can find bases among the subsets of these generating sets. First, we shall look for linearly independent subsets of $\{\mathscr{S}(t(i_1, \dots, i_p))\}$. In fact, we can prove that the set of distinct elements among them is linearly independent. By definition,

$$\begin{aligned}\mathscr{S}(t(i_1, \dots, i_p)) &= \frac{1}{p!} \sum_{\sigma \in \mathfrak{S}_p} t(i_{\sigma^{-1}(1)}, \dots, i_{\sigma^{-1}(p)}) \\ &= \frac{1}{p!} \sum_{\tau \in \mathfrak{S}_p} t(i_{\tau(1)}, \dots, i_{\tau(p)}).\end{aligned}$$

Since the sequence $(i_{\tau(1)}, \dots, i_{\tau(p)})$ is obtained by permuting $(i_1, \dots, i_p)$, the letters of $(i_{\tau(1)}, \dots, i_{\tau(p)})$ coincide with those of $(i_1, \dots, i_p)$ with their multiplicities (= the numbers of repetitions of a letter). Therefore, every term in $\mathscr{S}(t(i_1, \dots, i_p))$ has this property. Conversely, if the letters of $(j_1, \dots, j_p)$ coincide with those of $(i_1, \dots, i_p)$ with their multiplicities, there exists a one-to-one correspondence $\nu \leftrightarrow \tilde{\nu}$ $(1 \leq \nu \leq p,\ 1 \leq \tilde{\nu} \leq p)$ such that $j_\nu = j_{\tilde{\nu}}$. Namely, if we write $\tilde{\nu} = \tau(\nu)$, $\tau \in \mathfrak{S}_p$ and $j_\nu = i_{\tau(\nu)}$ $(1 \leq \nu \leq p)$. Moreover $\mathscr{S}(t(j_1, \dots, j_p)) = \mathscr{S}(t(i_1, \dots, i_p))$. In fact, since $t(j_1, \dots, j_p) = t(i_{\tau(1)}, \dots, i_{\tau(p)}) = P_{\tau^{-1}}(t(i_1, \dots, i_p))$, we have

$$\begin{aligned}\mathscr{S}(t(j_1, \dots, j_p)) &= \mathscr{S} P_\tau^{-1}(t(i_1, \dots, i_p)) \\ &= \mathscr{S}(t(i_1, \dots, i_p)).\end{aligned}$$

From above consideration, we can conclude that, if $\mathscr{S}(t(i_1, \dots, i_p)) \neq \mathscr{S}(t(i'_1, \dots, i'_p))$, they have no common terms, when they are written as a linear combination of standard basis elements. And we know that the set of distinct elements among $\mathscr{S}(t(i_1, \dots, i_p))$ is linearly independent and this set forms a basis for S^p. In particular, the dimension of S^p, which is the number of distinct elements among $\mathscr{S}(t(i_1, \dots, i_p))$, is equal to the number $\binom{n+p-1}{p}$ of p-combinations of $\{1, 2, \dots, n\}$ with repetitions (i.e., the ways of choosing p elements of $\{1, 2, \ldots, n\}$, disregarding order and allowing repeated elements). As representative of a p-combination, we can choose $(i_1, \ldots, i_p)$ such that $i_1 \leq i_2 \leq \cdots \leq i_p$. Thus we have the following proposition.

PROPOSITION 2.6. *$\{\mathscr{S}(t(i_1, \ldots, i_p)) | i_1 \leq i_2 \leq \cdots \leq i_p\}$ is a basis of the symmetric tensor space S^p of degree p. The dimension of S^p is given by*

$$\dim S^p = \binom{n+p-1}{p} = \frac{(n+p-1)!}{(n-1)!p!} \quad (\text{where } n = \dim V).$$

Next, let us consider the subspace A^p. We easily obtain the following lemma.

LEMMA 2.2. *Let $t(i_1, \ldots, i_p)$ be a standard basis element. If $i_k = i_l$ for some $k < l$ $(1 \leq k < l \leq p)$, $\mathscr{A}(t(i_1, \ldots, i_p)) = 0$.*

PROOF. Let τ be a transposition $(k\ l)$. From the assumption that $i_k = i_l$ we have

$$P_\tau(t(i_1, \ldots, i_p)) = t(i_1, \ldots, i_p).$$

Applying $\mathscr{A}$ to both sides of this equality, we have

$$\begin{aligned} \mathscr{A}(t(i_1, \ldots, i_p)) &= \mathscr{A} P_\tau(t(i_1, \ldots, i_p)) \\ &= \operatorname{sgn}(\tau)\mathscr{A}(t(i_1, \ldots, i_p)) \\ &= -\mathscr{A}(t(i_1, \ldots, i_p)), \end{aligned}$$

which shows $\mathscr{A}(t(i_1, \ldots, i_p)) = 0$. □

COROLLARY. *If $p > n = \dim V$, $\dim A^p = 0$.*

Therefore we consider the case $p \leq n$. By definition, we have

$$\mathscr{A}(t(i_1, \ldots, i_p)) = \frac{1}{p!} \sum_{\tau \in \mathfrak{S}_p} \operatorname{sgn}(\tau) t(i_{\tau(1)}, \ldots, i_{\tau(p)})$$

and we can proceed in a similar fashion to S^p. By Lemma 2.2, it is necessary to consider only these $t(i_1, \ldots, i_p)$, where $i_1, \ldots, i_p$ are all distinct. And, for these elements, $t(i_{\tau(1)}, \ldots, i_{\tau(p)}) \neq t(i_{\tau'(1)}, \ldots, i_{\tau'(p)})$, if $\tau \neq \tau'$. Thus we have $\mathscr{A}(t(i_1, \ldots, i_p) \neq 0$. Moreover, if p-element subsets $\{i_1, \ldots, i_p\}$ and $\{k_1, \ldots, k_p\}$ are equal as sets, we have

$$\mathscr{A}(t(i_1, \ldots, i_p)) = \pm\mathscr{A}(t(k_1, \ldots, k_p)).$$

And if these subsets are different, there are no common terms when we write $\mathscr{A}(t(i_1, \ldots, i_p))$ and $\mathscr{A}(t(k_1, \ldots, k_p))$ as linear combinations of standard basis elements $t(j_1, \ldots, j_p)$. Therefore, the set of all these $\mathscr{A}(t(i_1, \ldots, i_p))$ is linearly independent. The cardinality of this set is equal to the number $\binom{n}{p}$ of p-subsets of $\{1, \ldots, n\}$.

PROPOSITION 2.7. *Assume* $p \le n = \dim V$. *Then* $\{\mathscr{A}(t(i_1, \ldots, i_p)) | i_1 < i_2 < \cdots < i_p\}$ *form a basis for the alternating tensor space* A^p *of degree* p. *The dimension of* A^p *is given by* $\dim A^p = \binom{n}{p}$.

In particular, if $p = n$, *then* $\dim A^n = 1$. *For* $p < n$, $\dim A^p = \dim A^{n-p}$.

(c) Kernels of the symmetrizer and the alternator. Next, let us consider the kernels of the symmetrizer $\mathscr{S} = \mathscr{S}_p$ and the alternator $\mathscr{A} = \mathscr{A}_p$. For $t \in T^p$, we have

$$t - \mathscr{S}_p(t) = \frac{1}{p!}(p!t - \sum_{\sigma \in \mathfrak{S}_p} P_\sigma(t)) = \frac{1}{p!} \sum_{\sigma \in \mathfrak{S}_p} (t - P_\sigma(t)).$$

Suppose that t is an element of the kernel $\operatorname{Ker}\mathscr{S}_p$ of $\mathscr{S}_p$ i.e., $\mathscr{S}_p(t) = 0$. The above formula implies that $t = \frac{1}{p!}\sum_{\sigma \in \mathfrak{S}_p}(t - P_\sigma(t))$. On the other hand, for $\sigma \in \mathfrak{S}_p$, $t \in T^p$, set $t' = t - P_\sigma(t)$. Then $\mathscr{S}_p(t') = \mathscr{S}_p(t) - \mathscr{S}_p(P_\sigma(t)) = 0$ i.e., $t' \in \operatorname{Ker}\mathscr{S}_p$. Thus, if we denote by M^p the subspace generated by $t - P_\sigma(t)$ for all $\sigma \in \mathfrak{S}_p$ and all $t \in T^p$, we have $\operatorname{Ker}\mathscr{S}_p = M^p$. Moreover, to define M^p, we can restrict the range of σ to certain subsets. Namely, we have the following result.

PROPOSITION 2.8. *Let* X *be a subset of* $\mathfrak{S}_p$, *which satisfies the following condition* (G).

(G) *All elements of* $\mathfrak{S}_p$ *can be expressed as a product of a finite number of elements of* X.

If we denote by $M_X{}^p$ *the subspace of* T^p *generated by* $\{t - P_\tau(t) | t \in T^p, \tau \in X\}$, *we have* $M_X{}^p = \operatorname{Ker}\mathscr{S}_p$.

PROOF. At first, we show that, if we apply P_τ $(\tau \in X)$ on an element of $M_X{}^p$, the result belongs to $M_X{}^p$. Let $\tau' \in X$. For a generator $t - P_{\tau'}(t)$ of $M_X{}^p$,

$$P_\tau(t - P_{\tau'}(t)) = \{P_{\tau'}(t) - P_\tau(P_{\tau'}(t))\} + (t - P_{\tau'}(t)) - (t - P_\tau(t)).$$

As $P_{\tau'}(t) \in T^p$, $P_\tau(t - P_{\tau'}(t)) \in M_X{}^p$. Thus $P_\tau(M_X{}^p) \subseteq M_X{}^p$.

To prove the proposition, it is enough to show $M_X{}^p = M^p$. Let $\sigma \in \mathfrak{S}_p$. We can show $t - P_\sigma(t) \in M_X{}^p$ by induction on m, where m is the number of factors when σ is expressed as a product of elements of X. If $\sigma = \tau\sigma'$, where $\tau \in X$ and σ' is a product of $(m-1)$ elements of X,

$$t - P_\sigma(t) = t - P_\tau P_{\sigma'}(t) = t - P_\tau(t) + P_\tau(t - P_{\sigma'}(t)).$$

By assumption $t - P_{\sigma'}(t) \in M_X{}^p$ and $P_\tau(t - P_{\sigma'}(t)) \in M_X{}^p$ by the remark above. □

It is well known that X_1 = the set of transpositions of $\mathfrak{S}_p$ and $X_2 = \{(1\ 2), (2\ 3), \ldots, (p-1\ p)\}$ are subsets which satisfy the condition (G).

In the case of the alternator $\mathscr{A}_p$, only the signature makes a difference. For $t \in T^p$, we have

$$t - \mathscr{A}_p(t) = \frac{1}{p!} \sum_{\sigma \in \mathfrak{S}_p} \{t - \operatorname{sgn}(\sigma) P_\sigma(t)\}.$$

If we denote by N^p the subspace generated by $t - \operatorname{sgn}(\sigma) P_\sigma(t)$ for all $\sigma \in \mathfrak{S}_p$, $t \in T^p$, the kernel $\operatorname{Ker} \mathscr{A}_p$ of $\mathscr{A}_p$ is equal to N^p. In the same way as in Proposition 2.8, for a subset X of $\mathfrak{S}_p$ which satisfies the condition (G), denote by $N_X{}^p$ the subspace generated by $\{t - \operatorname{sgn}(\tau) P_\tau(t) | t \in T^p, \tau \in X\}$. Then we have $N_X{}^p = N^p$. For the proof, we use the following two equalities.

$$\begin{aligned} P_\tau(t - \operatorname{sgn}(\tau') P_{\tau'}(t)) &= \operatorname{sgn}(\tau)\operatorname{sgn}(\tau')\{P_{\tau'}(t) - \operatorname{sgn}(\tau) P_\tau(P_{\tau'}(t))\} \\ &\quad + \operatorname{sgn}(\tau)\{t - \operatorname{sgn}(\tau') P_{\tau'}(t)\} \\ &\quad - \operatorname{sgn}(\tau)\{t - \operatorname{sgn}(\tau) P_\tau(t)\} \qquad (\tau, \tau' \in X). \end{aligned}$$

Let $\sigma \in \mathfrak{S}_p$. If $\sigma = \tau\sigma'$, $\tau \in X$, then

$$\begin{aligned} t - \operatorname{sgn}(\sigma) P_\sigma(t) &= t - \operatorname{sgn}(\tau)\operatorname{sgn}(\sigma') P_{\tau\sigma'}(t) \\ &= \{t - \operatorname{sgn}(\tau) P_\tau(t)\} + \operatorname{sgn}(\tau) P_\tau(t - \operatorname{sgn}(\sigma') P_{\sigma'}(t)). \end{aligned}$$

Applying the above remark to $X = X_1$, the set of transpositions, we get the following proposition.

PROPOSITION 2.9. *Let N' be the subspace of T^p generated by all elements of the form $v_1 \otimes \cdots \otimes v_p$ such that $v_i = v_j$ for at least one pair $i \neq j$. Then N' is equal to* $\operatorname{Ker} \mathscr{A}_p$.

PROOF. Since $\operatorname{Ker} \mathscr{A}_p = N_{X_1}{}^p$, it is enough to show that the generators of N' are contained in $N_{X_1}{}^p$ and vice versa.

Let $t + P_\tau(t)$ $(t \in T^p, \tau \in X_1)$ be a generator of $N_{X_1}{}^p$. Since P_τ is linear, we may assume that t is of the form $t = v_1 \otimes \cdots \otimes v_p$. For a transposition $\tau = (i\ j)$ $(i < j)$

$$\begin{aligned} t + P_\tau(t) &= v_1 \otimes \cdots \otimes v_i \otimes \cdots \otimes v_j \otimes \cdots \otimes v_p \\ &\quad + v_1 \otimes \cdots \otimes v_j \otimes \cdots \otimes v_i \otimes \cdots \otimes v_p \\ &= v_1 \otimes \cdots \otimes (v_i + v_j) \otimes \cdots \otimes (v_i + v_j) \otimes \cdots \otimes v_p \\ &\quad - v_1 \otimes \cdots \otimes v_i \otimes \cdots \otimes v_i \otimes \cdots \otimes v_p \\ &\quad - v_1 \otimes \cdots \otimes v_j \otimes \cdots \otimes v_j \otimes \cdots \otimes v_p \in N'. \end{aligned}$$

Conversely, let $t' = v_1 \otimes \cdots \otimes v_p$ be a generator of N', i.e., $v_i = v_j$ for some $i \neq j$. Set $t = \frac{1}{2}t'$. Then $t - \operatorname{sgn}(\tau) P_\tau(t) = t + P_\tau(t) = t'$ for $\tau = (i\ j)$, which implies $t' \in N_{X_1}{}^p$. □

(d) Characterization by universality. From the definition of tensor product, $T^p(V)$ is characterized as the vector space which linearizes all p-multilinear mappings $\Phi: V \times \cdots \times V \to U$ (set $V_1 = \cdots = V_n = V$ in Theorem 1.1′). We mention analogous theorems for $S^p(V)$ and $A^p(V)$.

DEFINITION 2.3. Let $\Phi \in \mathscr{L}(V, \ldots, V; U)$ be a p-multilinear mapping.

(1) If the values are invariant for all permutations of variables, namely, for all $\sigma \in \mathfrak{S}_p$ and $v_i \in V$ $(i = 1, \ldots, p)$,
$$\Phi(v_{\sigma(1)}, \ldots, v_{\sigma(p)}) = \Phi(v_1, \ldots, v_p),$$
Φ is said to be *symmetric*.

(2) If $\Phi(v_1, \ldots, v_p) = 0$ whenever $v_i = v_j$ for some $i \neq j$, Φ is said to be *alternating*.

REMARK. If Φ is alternating, then for any $\sigma \in \mathfrak{S}_p$ and $v_i \in V$ $(i = 1, \ldots, p)$,
$$\Phi(v_{\sigma(1)}, \ldots, v_{\sigma(p)}) = \operatorname{sgn}(\sigma)\Phi(v_1, \ldots, v_p). \tag{2.7}$$
In fact, if Φ is alternating, for any transposition $\sigma = (i\ j)$ $(i < j)$ we have
$$\Phi(v_{\sigma(1)}, \ldots, v_{\sigma(p)}) = \operatorname{sgn}(\sigma)\Phi(v_1, \ldots, v_p),$$
i.e., $\Phi(v_1, \ldots, v_j, \ldots, v_i, \ldots, v_p) = -\Phi(v_1, \ldots, v_i, \ldots, v_j, \ldots, v_p)$. From this, the general result follows. Now, for $\sigma = (i\ j)$, by definition
$$\Phi\Big(v_1, \ldots, \underbrace{v_i + v_j}_{i\text{th position}}, \ldots, \underbrace{v_i + v_j}_{j\text{th position}}, \ldots, v_p\Big) = 0.$$
Expand this using multilinearlity, and substitute
$$\Phi(v_1, \ldots, v_i, \ldots, v_i, \ldots, v_p) = 0$$
and
$$\Phi(v_1, \ldots, v_j, \ldots, v_j, \ldots, v_p) = 0.$$
Then we have the above equality. Conversely, (precisely speaking, if the characteristic of k is not equal to 2), assume that Φ satisfies (2.7) and $v_i = v_j$ for some $i \neq j$. Then (2.7) for $\sigma = (i\ j)$ implies $\Phi(v_1, \ldots, v_p) = -\Phi(v_1, \ldots, v_p)$, i.e., $\Phi(v_1, \ldots, v_p) = 0$. That is, if Φ satisfies (2.7), we know that Φ is alternating.

EXAMPLE 2.12. Let V be the k-vector space of $n \times n$ matrices with coefficients in k. Define a mapping $\Phi: V \times V \to V$ by
$$\Phi(A, B) = AB - BA \qquad (A, B \in V).$$
Then Φ is an alternating bilinear mapping.

EXAMPLE 2.13. If $U = k$, a symmetric (resp. alternating) p-multilinear mapping $\in \mathscr{L}(V, \ldots, V; k)$ is called a symmetric (resp. alternating) p-multilinear form. Let $V = k^n$, the space of column vectors. Define Φ: $V \times \cdots \times V \to k$ by
$$\Phi(v_1, \ldots, v_n) = \det(A) \qquad (v_i \in V),$$

where A is the $n \times n$ matrix whose ith column is v_i $(i = 1, \ldots, n)$. Then, as is well known, Φ is an alternating n-multilinear form.

EXAMPLE 2.14. Let $\iota: V \times \cdots \times V$ (p factors) $\to T^p(V)$ be the canonical mapping of tensor product $T^p(V) = V \otimes \cdots \otimes V$. The composite mappings

$$\mathscr{S}_p \circ \iota: V \times \cdots \times V \to S^p(V),$$
$$\mathscr{A}_p \circ \iota: V \times \cdots \times V \to A^p(V)$$

are a symmetric and alternating p-multilinear mapping respectively.

In fact, these mappings have a *universal property*. We have the following theorem.

THEOREM 2.1. (1) *Let V and U be k-vector spaces. If $\Phi \in \mathscr{L}(V, \ldots, V; U)$ is a symmetric p-multilinear mapping, then there exists a unique linear mapping $F: S^p(V) \to U$ such that $F \circ (\mathscr{S}_p \circ \iota) = \Phi$.*

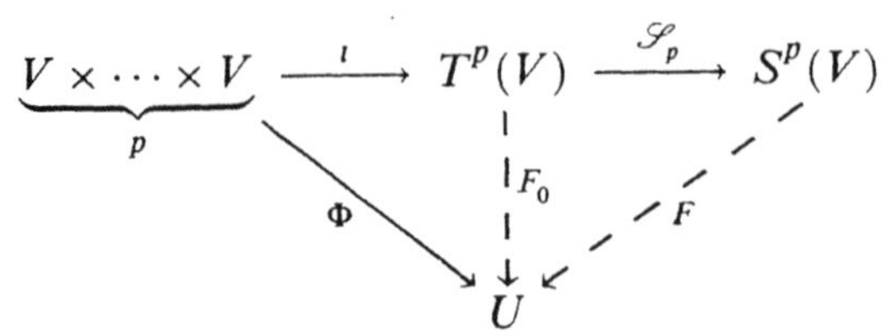

(2) *A similar result holds for alternating p-multilinear mappings. That is, if $\Psi \in \mathscr{L}(V, \ldots, V; U)$ is an alternating p-multilinear mapping, then there exists a unique linear mapping $G: A^p(V) \to U$ such that $G \circ (\mathscr{A}_p \circ \iota) = \Psi$.*

PROOF. (1) By the fundamental property (T) of tensor products, Φ induces a unique linear mapping $F_0: T^p(V) \to U$. Suppose that there exists a linear mapping $F: S^p(V) \to U$ which satisfies the condition of the theorem, then $F \circ \mathscr{S}_p$ satisfies the requirement for F_0. Thus the uniqueness of F_0 implies $F \circ \mathscr{S}_p = F_0$. For $x \in S^p(V)$, $F(x) = F_0(x)$ is necessary since $\mathscr{S}_p(x) = x$. In particular, this shows the uniqueness of F.

Now we define a mapping $F: S^p(V) \to U$ by $F(x) = F_p(x)$ for $x \in S^p(V)$. Since F_0 is linear, F is linear also. It is necessary to show that $F \circ (\mathscr{S}_p \circ \iota) = \Phi$.

For any $t \in T^p(V)$, $\mathscr{S}_p(t) \in S^p(V)$. Therefore, by definition of F we have $(F \circ \mathscr{S}_p)(t) = (F_0 \circ \mathscr{S}_p)(t)$ for $t \in T^p(V)$. So, if we show that $F_0 \circ \mathscr{S}_p = F_0$, we have $F \circ \mathscr{S}_p = F_0$ and $F \circ (\mathscr{S}_p \circ \iota) = F_0 \circ \iota = \Phi$. By linearlity of F_0 and $\mathscr{S}_p$, it is enough to show $(F_0 \circ \mathscr{S}_p)(t) = F_0(t)$ for $t = v_1 \otimes \cdots \otimes v_p$ $(v_i \in V)$. On the other hand, for $\sigma \in \mathscr{S}_p$

$$\begin{aligned} F_0 \circ P_\sigma(v_1 \otimes \cdots \otimes v_p) &= F_0(v_{\sigma^{-1}(1)} \otimes \cdots \otimes v_{\sigma^{-1}(p)}) \\ &= F_0 \circ \iota(v_{\sigma^{-1}(1)}, \ldots, v_{\sigma^{-1}(p)}) \\ &= \Phi(v_{\sigma^{-1}(1)}, \ldots, v_{\sigma^{-1}(p)}) \\ &= \Phi(v_1, \ldots, v_p) \quad \text{(since } \Phi \text{ is symmetric)}. \end{aligned}$$

Thus, for $t = v_1 \otimes \cdots \otimes v_p$, we have

$$\begin{aligned} F_0 \circ \mathscr{S}_p(t) &= \frac{1}{p!} \sum_{\sigma \in \mathfrak{S}_p} F_0 \circ P_\sigma(v_1 \otimes \cdots \otimes v_p) \\ &= \Phi(v_1, \dots, v_p) = F_0 \circ \iota(v_1, \dots, v_p) \\ &= F_0(v_1 \otimes \cdots \otimes v_p) = F_0(t). \end{aligned}$$

(2) is proved in the same way. □

(e) Case of covariant tensors. As for the space $T_p(V) = T^p(V^*)$ of covariant tensors of degree p, we can define a linear transformation P_σ^* of $T_p(V)$ corresponding to an element σ of $\mathfrak{S}_p$, in a similar fashion to contravariant tensor's case. Namely, P_σ^* is defined by

$$P_\sigma^*(\varphi_1 \otimes \cdots \otimes \varphi_p) = \varphi_{\sigma^{-1}(1)} \otimes \cdots \otimes \varphi_{\sigma^{-1}(p)}$$

for $\varphi_1, \dots, \varphi_p \in V^*$. Using P_σ^*, symmetric tensors, alternating tensors, and the subspaces $S^p(V^*)$, $A^p(V^*)$ consisting of them are defined.

As we have already discussed in Examples 2.2 and 2.3 §1, $T_p(V)$ can be identified with the dual space of $T^p(V)$ and $T_p(V) \cong \mathscr{L}(V, \dots, V; k)$ (= the space of p-multilinear forms on V). When $T_p(V)$ is identified with the dual space of $T^p(V)$, $P_\sigma^*(\varphi_1 \otimes \cdots \otimes \varphi_p)$ is a linear form on $T^p(V)$ and

$$\begin{aligned} P_\sigma^*(\varphi_1 \otimes \cdots \otimes \varphi_p)(v_1 \otimes \cdots \otimes v_p) &= (\varphi_{\sigma^{-1}(1)} \otimes \cdots \otimes \varphi_{\sigma^{-1}(p)})(v_1 \otimes \cdots \otimes v_p) \\ &= \varphi_{\sigma^{-1}(1)}(v_1) \cdots \varphi_{\sigma^{-1}(p)}(v_p) \\ &= \varphi_1(v_{\sigma(1)}) \cdots \varphi_p(v_{\sigma(p)}) \\ &= (\varphi_1 \otimes \cdots \otimes \varphi_p)(v_{\sigma(1)} \otimes \cdots \otimes v_{\sigma(p)}) \\ &= (\varphi_1 \otimes \cdots \otimes \varphi_p)(P_\sigma^{-1})(v_1 \otimes \cdots \otimes v_p) \end{aligned}$$

$$(\varphi_i \in V^*, \ v_j \in V).$$

This equality shows that $P_\sigma^* = {}^t P_\sigma^{-1}$.

Next, considering the isomorphism $T_p(V) \cong \mathscr{L}(V, \dots, V; k)$, we denote by $\Phi(t^*)$ the p-multilinear form on V corresponding to $t^* \in T_p(V) = (T^p(V))^*$. That is,

$$\Phi(t^*)(v_1, \dots, v_p) = t^*(v_1 \otimes \cdots \otimes v_p)$$
$$(v_i \in V, \ t^* \in (T^p(v))^*).$$

Then the p-multilinear form $\Phi(P_\sigma^*(t^*))$ is obtained by permuting the variables of $\Phi(t^*)$, as we can see by

$$\begin{aligned}\Phi(P_\sigma^*(t^*))(v_1, \dots, v_p) &= P_\sigma^*(t^*)(v_1 \otimes \cdots \otimes v_p) \\ &= t^*(P_\sigma^{-1}(v_1 \otimes \cdots \otimes v_p)) \\ &= t^*(v_{\sigma(1)} \otimes \cdots \otimes v_{\sigma(p)}) \\ &= \Phi(t^*)(v_{\sigma(1)}, \dots, v_{\sigma(p)}).\end{aligned}$$

In particular, the p-multilinear forms corresponding to $t^* \in S^p(V^*)$ (resp. $A^p(V^*)$) are symmetric (resp. alternating).

§4. Tensor algebras and their properties

(a) Definition of tensor algebra. Collecting all contravariant tensors of different degree, we define the (infinite) direct sum of $T^p(V)$, $p = 0, 1, \dots$,

$$T(V) = \bigoplus_{p=0}^{\infty} T^p(V).$$

In general, suppose that, to each element i of $I = \{0, 1, \dots, \}$ there corresponds a k-vector space V_i. Then we define the *infinite direct sum* $\widetilde{V} = \bigoplus_{i=0}^{\infty} V_i$ as follows. As a set, $\widetilde{V}$ consists of all infinite sequences

$$v = (v_0, v_1, \dots) \qquad (v_i \in V_i)$$

such that the number of indices i with $v_i \neq 0$ is finite. If we define, for $v = (v_0, v_1, \dots) \in \widetilde{V}$ and $v' = (v_0', v_1', \dots) \in \widetilde{V}$, the sum $v + v'$ and the scalar multiple αv $(\alpha \in k)$ by the following formulas, then $\widetilde{V}$ has the structure of a k-vector space (not necessarily finite dimensional).

$$v + v' = (v_0 + v_0', v_1 + v_1', \dots), \qquad \alpha v = (\alpha v_0, \alpha v_1, \dots).$$

If we identify $v_i \in V_i$ with the element

$$(0, \dots, 0, \underset{i\text{th term}}{v_i}, 0, \dots)$$

of $\widetilde{V}$, then V_i can be considered as a subset of $\widetilde{V}$. Let $v = (v_0, v_1, \dots) \in \widetilde{V}$. Then there exists an $i_0 \in I$ such that $v_j = 0$ for all $j > i_0$. Thus we can write

$$v = \sum_{j=0}^{i_0} v_j \qquad (v_j \in V_j).$$

We write this expression formally as $v = \sum_{j=0}^{\infty} v_j$.

By setting $V_i = T^i(V)$ in the definition of (infinite) direct sum, the (infinite) direct sum $T(V)$ of contravariant tensor spaces $T^i(V)$ becomes a k-vector space (in general, infinite dimensional).

Next, let us define the product of two elements of $T(V)$. As we explained in §2 (b), using the tensor product of $T^p(V)$ and $T^q(V)$, the product of

$x \in T^p(V)$ and $y \in T^q(V)$ is defined and belongs to $T^{p+q}(V)$. Denote it by $x \otimes y$. Moreover, the mapping

$$\begin{array}{ccc} T^p(V) \times T^q(V) & \to & T^{p+q}(V) \\ \Psi & & \Psi \\ (x, y) & \mapsto & x \otimes y \end{array}$$

is bilinear. Thus, for $t = \sum_{p=0}^{\infty} t_p$, $t' = \sum_{q=0}^{\infty} t'_q$, $t_p, t'_p \in T^p(V)$, define $t \otimes t'$ as follows;

$$t \otimes t' = \sum_{p=0}^{\infty} \left(\sum_{r+s=p} t_r \otimes t'_s \right).$$

Clearly $t \otimes t' \in T(V)$ and it is called the product of t and t'. The mapping

$$\begin{array}{ccc} T(V) \times T(V) & \to & T(V) \\ \Psi & & \Psi \\ (t, t') & \mapsto & t \otimes t' \end{array}$$

is bilinear. From the associativity of tensor product, the multiplication thus defined satisfies the associativity law, i.e., for $t, t', t'' \in T(V)$,

$$(t \otimes t') \otimes t'' = t \otimes (t' \otimes t'').$$

If we consider $1 \in k = T^0(V)$ as an element of $T(V)$, we have, for all $t \in T$, $t \otimes 1 = 1 \otimes t = t$.

The vector space $T(V)$ equipped with this multiplication is called the *tensor algebra* of V. When we regard $T^p(V)$ as a subspace of $T(V)$, the element of the form $v_1 \otimes \cdots \otimes v_p$ $(v_i \in V)$ of $T^p(V)$ is called a *decomposable tensor.*

(b) Preliminaries for associative algebras. Here we define several notions which we need later.

Let A be a k-vector space (not necessarily finite dimensional). If A has a bilinear multiplication which is associative, that is if there exists a bilinear mapping $m: A \times A \to A$ such that m satisfies associativity (§1.10, Remark), then A (more precisely the pair (A, m)) is called an *associative algebra* over k. Usually $m(a, a')$ is written as aa' or $a \cdot a'$. If there exists a unit element e for the multiplication, e is called the *unit element* of A.

EXAMPLE 2.15. Let $A = M_n(k)$, the k-vector space of $n \times n$ matrices with coefficients in k, and $m(a, a') = aa'$ (product of matrices). Then A is an associative algebra.

EXAMPLE 2.15′. Let V be a k-vector space. Set $A = \mathrm{Hom}(V, V)$, the k-vector space of linear transformations of V, and $m(a, a') = a \circ a'$ (composition of mappings). Then A is an associative algebra.

EXAMPLE 2.16. Let V be a k-vector space. The tensor algebra $T(V)$ with $m(t, t') = t \otimes t'$ is an associative algebra.

Let (A, m) and (A', m') be associative algebras. If a linear mapping Φ of a k-vector space A to A' satisfies $\Phi \circ m = m' \circ \Phi$, that is, if

$\Phi(aa') = \Phi(a)\Phi(a')$ for all $a, a' \in A$, then Φ is called an *associative algebra homomorphism*. Moreover, if A and A' have the unit elements e and e' respectively, an associative algebra homomorphism must satisfy also $\Phi(e) = e'$. In particular, if Φ is one-to-one and onto, Φ is called an (associative algebra) isomorphism. When there exists an isomorphism $\Phi: A \to A'$, A and A' are said to be isomorphic.

Let (A, m) be an associative algebra. If a (vector) subspace B of A satisfies the condition $m(B, B) \subseteq B$, B is called a *subalgebra* of A. When B is a subalgebra, B is regarded as an associative algebra with the restriction of m to $B \times B$.

If a subalgebra $\mathfrak{a}$ of an associative algebra (A, m) satisfies the conditions $m(A, \mathfrak{a}) \subseteq \mathfrak{a}$ and $m(\mathfrak{a}, A) \subseteq \mathfrak{a}$, then $\mathfrak{a}$ is called an *ideal* of A.

Let $\mathfrak{a}$ be an ideal of A. Construct the factor space $A/\mathfrak{a}$ of the k-vector space A with a subspace $\mathfrak{a}$. We can define multiplication $\overline{m}$ of $A/\mathfrak{a}$ as follows: For $A/\mathfrak{a} \ni \overline{a} = a + \mathfrak{a}$ and $\overline{a}' = a' + \mathfrak{a}$,

$$\overline{m}(\overline{a}, \overline{a}') = \overline{m(a, a')} \qquad (= \text{the coset of } m(a, a')).$$

Since $\mathfrak{a}$ is an ideal, $\overline{m}(\overline{a}, \overline{a}') \in A/\mathfrak{a}$ is independent of the choice of coset representatives for $\mathfrak{a}$ and $\mathfrak{a}'$. Thus we get a bilinear mapping $\overline{m}: A/\mathfrak{a} \times A/\mathfrak{a} \to A/\mathfrak{a}$ and $(A/\mathfrak{a}, \overline{m})$ is an associative algebra. This is called a *factor algebra* of A by $\mathfrak{a}$.

EXAMPLE 2.17. Let (A, m) be an associative algebra and let $\{B_i\}_{i\in I}$ be a family of subalgebras. Then the intersection $\bigcap_{i\in I} B_i$ is also a subalgebra.

Let S be a subset of A. The intersection of all subalgebras which contain S is called the *subalgebra generated by* S and denoted by B_S. B_S is the smallest subalgebra containing S.

If $S = \{s\}$, we can easily verify that $B_S = \{\sum_{i=1}^n \alpha_i s^i | n = 1, 2, \dots, \alpha_i \in k\}$. In general, make all possible finite products $s_i \cdots s_j$ (some elements can appear twice or more) of elements of S, then B_S is the set of all finite linear combinations of these with coefficients in k.

In particular, if $B_S = A$, we say that S is a *set of generators* for A or that A is generated by S.

As sets of generators for tensor algebra $T(V)$, we can take, for example, $S_0 = \{1\} \cup V$, $S_1 = \{1\} \cup \{e_1, \dots, e_n\}$, where $(e_1, \dots, e_n)$ is a basis for V.

EXAMPLE 2.18. Let (A, m) be an associative algebra and $\{\mathfrak{a}_i\}_{\lambda\in I}$ a family of ideals. Then the intersection $\bigcap_{i\in I} \mathfrak{a}_i$ is also an ideal.

Let S be a subset of A. The intersection of all ideals which contain S is called the *ideal generated by* S and denoted by $\mathfrak{a}_S$. $\mathfrak{a}_S$ is the set of all (finite) linear combinations with coefficients in k of those elements which are obtained by multiplying elements of A to left or right or both sides of elements of S.

Next, we shall define a grading. When, to each element i of $I = \{0, 1, \dots\}$, there corresponds a k-vector space A_i and A is the (infinite) direct sum $A = \bigoplus_{i=0}^{\infty} A_i$ of them, then A is called a *graded vector space* (of type

I). ([3]) Then we can view $A_i \subseteq A$. Elements of some A_i are called *homogeneous elements* and the index i such that $a \in A_i$ is called the *degree* of the homogeneous element a. Moreover, if $a \in A$ is written as $a = \sum_{i=0}^{\infty} a_i$ ($a_i \in A_i$), a_i is called the *homogeneous component* of a of degree i.

If a graded vector space $A = \bigoplus_{i=0}^{\infty} A_i$ is an associative algebra and the multiplication m, for all i and j, satisfies

$$m(A_i, A_j) \subseteq A_{i+j},$$

then A is called a *graded associative algebra.* The tensor algebra $T(V) = \bigoplus_{p=0}^{\infty} T^p(V)$ is an example of graded associative algebra.

Let $A = \bigoplus_{i=0}^{\infty} A_i$ be a graded associative algebra and B a subalgebra of A. B is called a *graded subalgebra* of A, if B can be expressed as the sum $B = \bigoplus_{i=0}^{\infty}(B \cap A_i)$. (That is, every element b of B can be written in the form $b = \sum_{i=0}^{\infty} b_i$, where $b_i \in B \cap A_i$.)

PROPOSITION 2.10. *Let $A = \bigoplus_{i=0}^{\infty} A_i$ be a graded associative algebra and B a subalgebra of A. The following three conditions are equivalent.*

(1) *B is a graded subalgebra of A.*

(2) *For all $b \in B$, the homogeneous components of b are also contained in B.*

(3) (*As associative algebra*) *B is generated by homogeneous elements.*

PROOF. Any element $b \in B$ can be expressed uniquely as $b = \sum a_i$ where $a_i \in A_i$. Thus (1) and (2) are equivalent. (1) $\Rightarrow$ (3) is obvious, too. To show (3) $\Rightarrow$ (2), let $\{b_\lambda\}$ $(\lambda \in \Lambda)$ be a set of generators for B consisting of homogeneous elements. Then, all $b \in B$ can be expressed as $b = \sum \alpha_{(\lambda_1, \dots, \lambda_t)} b_{\lambda_1} \cdots b_{\lambda_t}$ $(\alpha_{(\lambda_1, \dots, \lambda_t)} \in k)$. The homogeneous component of b of degree i is the sum of all the terms $\alpha_{(\lambda_1, \dots, \lambda_t)} b_{\lambda_1} \cdots b_{\lambda_t}$, whose degrees are equal to i. This shows (2). □

An ideal $\mathfrak{a}$ of graded associative algebra is called a *graded ideal* or a *homogeneous ideal* if it is graded as subalgebra. Similarly to the case of graded subalgebras, we have the following proposition.

PROPOSITION 2.11. *Let A be a graded associative algebra and let $\mathfrak{a}$ be an ideal of A. The following three conditions are equivalent.*

(1) *$\mathfrak{a}$ is a graded ideal of A.*

(2) *For all $a \in \mathfrak{a}$, the homogeneous components of a are also contained in $\mathfrak{a}$.*

(3) *$\mathfrak{a}$ is generated as an ideal by homogeneous elements.*

PROOF. Only the assertion (3) $\Rightarrow$ (2) is necessary to prove. Let $\{x_\lambda\}$ $(\lambda \in \Lambda)$ be the set of generators for $\mathfrak{a}$ consisting of homogeneous elements. All elements x of $\mathfrak{a}$ can be expressed in the form $x = \sum_\lambda a_\lambda x_\lambda b_\lambda$ $(a_\lambda, b_\lambda \in A)$

([3]) In general, for the set I of grading, we can take a commutative semigroup with a unit element (where the operation is denoted by $+$).

(where some terms may not have the parts a_λ or b_λ). Let

$$a_\lambda = \sum_{i=0}^{\infty} a_{\lambda,i}, \qquad b_\lambda = \sum_{i=0}^{\infty} b_{\lambda,i}$$

be the decomposition of a_λ and b_λ into homogeneous components. Then

$$x = \sum_\lambda a_\lambda x_\lambda b_\lambda = \sum_{k=0}^{\infty} \left(\sum_{i+\delta(\lambda)+j=k} a_{\lambda,i} x_\lambda b_{\lambda,j} \right).$$

(We denote by $\delta(\lambda)$ the degree of x_λ.) The homogeneous components of x are $\sum a_{\lambda,i} x_\lambda b_{\lambda,j}$, which are elements of $\mathfrak{a}$. □

If $\mathfrak{a}$ is a graded ideal, $\mathfrak{a}$ can be written as $\mathfrak{a} = \bigoplus_{i=0}^{\infty} \mathfrak{a}_i$, where $\mathfrak{a}_i = \mathfrak{a} \cap A_i$. Thus the factor algebra $A/\mathfrak{a}$ can be considered as a direct sum of $A_i/\mathfrak{a}_i$ $(i \in I)$. Moreover, since $m(A_i, \mathfrak{a}_j) \subseteq a_{i+j}$ and $m(\mathfrak{a}_i, A_j) \subseteq \mathfrak{a}_{i+j}$, where m is the multiplication on A, the induced multiplication $\overline{m}$ of $A/\mathfrak{a}$ satisfies

$$\overline{m}(A_i/\mathfrak{a}_i, A_j/\mathfrak{a}_j) \subseteq A_{i+j}/\mathfrak{a}_{i+j}.$$

Thus $A/\mathfrak{a} = \bigoplus_{i=0}^{\infty}(A_i/\mathfrak{a}_i)$ is naturally a graded associative algebra.

(c) Universality of tensor algebra. Coming back to the main theme of the book, we describe the *universality* of the tensor algebra $T(V)$. We have the following theorem.

THEOREM 2.2. *Let V be a k-vector space, A an associative algebra with the unit element e, and F a linear mapping of V into A. Then F can be extended uniquely to an associative algebra homomorphism $\widetilde{F}: T(V) \to A$. Namely, there exists a unique associative algebra homomorphism $\widetilde{F}$ such that $\widetilde{F}(1) = e$ and $\widetilde{F} \circ \iota = F$, where ι denotes the natural inclusion mapping of $V = T^1(V)$ into $T(V)$.*

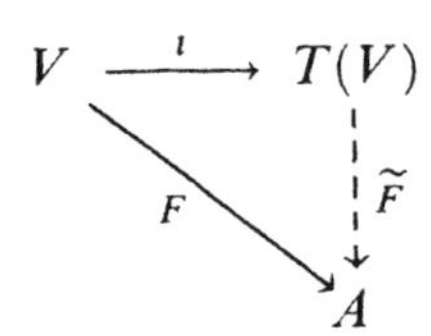

PROOF. For all p $(p \geq 2)$, define a p-multilinear mapping

$$\Phi_p: \underbrace{V \times \cdots \times V}_{p \text{ factors}} \to A, \qquad \Phi_p(v_1, \ldots, v_p) = F(v_1) \cdots F(v_p).$$

By the fundamental property (T) of tensor products (Theorem 1.1′), Φ_p induces a linear mapping

$$F_p: V \otimes \cdots \otimes V = T^p(V) \to A,$$
$$F_p(v_1 \otimes \cdots \otimes v_p) = F(v_1) \cdots F(v_p).$$

Moreover, defining $F_1 = F$ and $F_0(\alpha) = \alpha e$ $(\alpha \in k)$, we have $F_1: V = T^1(V) \to A$ and $F_0: k = T^0(V) \to A$. Then, assigning $\widetilde{F}(t) = \sum_p F_p(t_p)$

to $T(V) \ni t = \sum_p t_p$ $(t_p \in T^p(V))$, we can define a linear mapping $\widetilde{F}$: $T(V) \to A$.

For decomposable tensors t and t', by definition we have

$$\widetilde{F}(t \otimes t') = \widetilde{F}(t)\widetilde{F}(t').$$

Since any element of $T(V)$ is a sum of decomposable tensors and $\widetilde{F}$ is a linear mapping, it follows that $\widetilde{F}$ is an associative algebra homomorphism. To show the uniqueness of $\widetilde{F}$, we remark that the values of $\widetilde{F}$ on the set $V \cup \{1\}$ of generators of $T(V)$ are determined by the conditions $\widetilde{F} \circ \iota = F$ and $\widetilde{F}(1) = e$. Since $\widetilde{F}$ is an associative algebra homomorphism, $\widetilde{F}$ is uniquely determined on $T(V)$ by F. □

COROLLARY. *Let V, W be k-vector spaces and F a linear mapping of V into W. Then F is uniquely extended to an associative algebra homomorphism $T(F)$ of $T(V)$ into $T(W)$. In particular, a linear transformation of V is extended to an (associative algebra) endomorphism of $T(V)$.*

PROOF. Let j be the natural inclusion mapping of $W = T^1(W)$ into $T(W)$. Apply the theorem to $j \circ F : V \to T(W)$. □

REMARK. In view of the proof of the theorem, for $v_1 \otimes \cdots \otimes v_p \in T^p(V)$, $T(F)$ is given by

$$T(F)(v_1 \otimes \cdots \otimes v_p) = F(v_1) \otimes \cdots \otimes F(v_p).$$

That is, $T(F)$ is given by the "sum" of tensor products $\underbrace{F \otimes \cdots \otimes F}_{p \text{ factors}}$: $T^p(V) \to T^p(W)$ of the linear mapping F (§1.7).

We have already used the word "universality" in connection with the definition of tensor product (§1.3). Here "universality of tensor algebra" means that, as explained in Theorem 2.2, the tensor algebra $T(V)$ and the inclusion mapping ι dominate, in a sense, the linear mappings of V into an associative algebra. Moreover, we have Theorem 2.3, which asserts that the tensor algebra is "characterized" by this property. In other words, the associative algebra, the existence and uniqueness of which are assured by Theorem 2.3, is a tensor algebra.

THEOREM 2.3. (1) *Let V be a k-vector space. There exist an associative algebra A_0 with unit element e_0 and a linear mapping ι_0 of V into A_0 which have the following property* (TA):

(TA) *Let F be a linear mapping of V into an associative algebra A_1 with the unit element e_1. Then there exists a unique associative algebra homomorphism $\widetilde{F} : A_0 \to A_1$ such that $\widetilde{F} \circ \iota_0 = F$,*

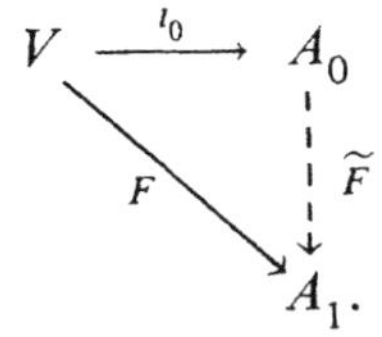

(2) *The pair* (A_0, ι_0) *is unique in the following sense: If the pairs* (A_0, ι_0) *and* (A_0', ι_0') *consisting of an associative algebra and a linear mapping have the property* (TA), *then there exists a unique associative algebra isomorphism* $F_0: A_0 \to A_0'$ *such that* $\tilde{F}_0 \circ \iota_0 = \iota_0'$.

PROOF. Part (1) is obtained by Theorem 2.2, which says that $A_0 = T(V)$ and $\iota_0 = \iota$ have the property (TA).

(2) Suppose that both (A_0, ι_0) and (A_0', ι_0') have the property (TA). By (TA) for (A_0, ι_0) there exists a unique homomorphism $\tilde{F}_0: A_0 \to A_0'$ such that $\tilde{F}_0 \circ \iota_0 = \iota_0'$. Similarly, we have a unique $\tilde{F}_0': A_0' \to A_0$ such that $\tilde{F}_0' \circ \iota_0' = \iota_0$,

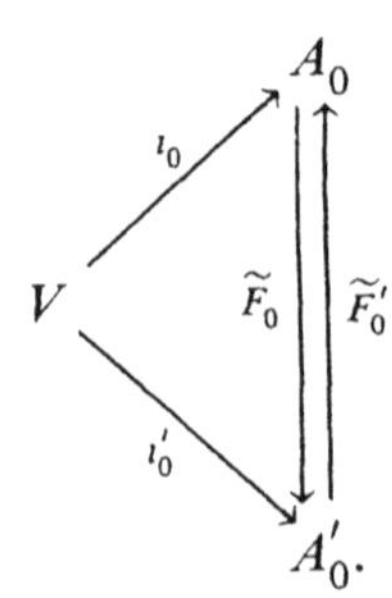

Substitute $\iota_0' = \tilde{F}_0 \circ \iota_0$ in the equality $\tilde{F}_0' \circ \iota_0' = \iota_0$. Then we have $\tilde{F}_0' \circ \tilde{F}_0 \circ \iota_0 = \iota_0$. On the other hand, the identity mapping id of A_0 clearly satisfies $\mathrm{id} \circ \iota_0 = \iota_0$. Thus, by the uniqueness assertion of (TA), we have $\tilde{F}_0' \circ \tilde{F}_0 = \mathrm{id}$. Similarly, $\tilde{F}_0 \circ \tilde{F}_0' =$ the identity mapping of A_0'. Thus $\tilde{F}_0$ is an isomorphism. □

COROLLARY. *Let* V *be a* k*-vector space of dimension* n *and* $(e_1, \dots, e_n)$ *a basis for* V. *Let* A *be an associative algebra generated by* n *elements* $\{a_1, \dots, a_n\}$ *and the unit element* e. *Then there exists a unique associative algebra homomorphism* $\tilde{F}: T(V) \to A$ *such that* $\tilde{F}(e_i) = a_i$ $(i = 1, \dots, n)$. *Clearly* $\tilde{F}$ *is surjective.*

PROOF. Since $(e_1, \dots, e_n)$ is a basis of V, a linear mapping $F: V \to A$ is defined by $F(e_i) = a_i$ $(i = 1, \dots, n)$. Applying (TA), we have the result. □

REMARK. The property of $T(V)$ mentioned in this corollary is sometimes referred to by saying " $T(V)$ is the *free associative algebra* generated by n elements" or " $T(V)$ is a *noncommutative polynomial algebra*."

(d) Derivations. In the corollary to Theorem 2.2 we observed that a linear transformation of V can be extended to an endomorphism of $T(V)$. Now we shall describe another extension. For this purpose, we give the following definition.

DEFINITION 2.4. Let A be a (not necessarily finite-dimensional) k-vector space and m a bilinear multiplication on A. Denote $m(a, a')$ by $a \cdot a'$ for a, $a' \in A$. A linear transformation D of A is called a *derivation* if, for all a, $a' \in A$, we have

$$D(a \cdot a') = D(a) \cdot a' + a \cdot D(a').$$

As we see from the above definition, the concept of derivation is a generalization of differentiation in calculus. Let e be the unit element of A. Then $D(e) = 0$, since $D(e \cdot e) = D(e) \cdot e + e \cdot D(e) = D(e) + D(e)$.

Once we define a derivation, we have the next proposition.

PROPOSITION 2.12. *Let V be a k-vector space and F a linear transformation of V. Then F can be uniquely extended to a derivation $D(F)$ of the tensor algebra $T(V)$.*

PROOF. If there exists a derivation $D(F)$ of $T(V)$ such that $D(F)|V$ (= the restriction of $D(F)$ to V) = F, then

$$\begin{aligned} D(F)(v_1 \otimes v_2) &= D(F)(v_1) \otimes v_2 + v_1 \otimes D(F)(v_2) \\ &= F(v_1) \otimes v_2 + v_1 \otimes F(v_2). \end{aligned}$$

More generally, we must have

$$D(F)(v_1 \otimes \cdots \otimes V_p) = \sum_{i=1}^{p} v_1 \otimes \cdots \otimes F(v_i) \otimes \cdots \otimes v_p.$$

Thus, in order to construct $D(F)$, we define for all p $(p \geq 2)$ a p-multilinear mapping $\Psi_p \colon V \times \cdots \times V \to T^p(V)$ as follows;

$$\Psi_p(v_1, \ldots, v_p) = \sum_{i=1}^{p} v_1 \otimes \cdots \otimes F(v_i) \otimes \cdots \otimes v_p.$$

By property (T) of tensor products, Ψ_p induces a linear mapping

$$\begin{aligned} &D(F)_p \colon T^p(V) \to T^p(V), \\ &D(F)_p(v_1 \otimes \cdots \otimes v_p) = \Psi_p(v_1, \ldots, v_p). \end{aligned}$$

Setting $D(F)_1 = F \colon V \to T(V)$ and $D(F)_0 = 0$ (zero-mapping): $k \to T(V)$, we define $D(F)(t) = \sum_p D(F)_p(t_p)$ for $T(V) \ni t = \sum_p t_p$ $(t_p \in T^p(V))$. Then $D(F) \colon T(V) \to T(V)$ is a linear mapping such that $D(F)|V = F$.

Next, we shall prove that $D(F)$ is a derivation. Namely, we shall show that, for t, $t \in T(V)$,

$$D(F)(t \otimes t') = D(F)(t) \otimes t' + t \otimes D(F)(t').$$

It suffices to show this when t and t' are decomposable. Let $t = v_1 \otimes \cdots \otimes v_p$ and $t' = v_{p+1} \otimes \cdots \otimes v_{p+q}$ $(v_i \in V)$. If $p \geq 1$ and $q \geq 1$, we have

$$\begin{aligned} D(F)(t \otimes t') &= D(F)(v_1 \otimes \cdots \otimes v_p \otimes v_{p+1} \otimes \cdots \otimes v_{p+q}) \\ &= \sum_{i=1}^{p+q} v_1 \otimes \cdots \otimes F(v_i) \otimes \cdots \otimes v_{p+q} \\ &= \sum_{i=1}^{p} v_1 \otimes \cdots \otimes F(v_i) \otimes \cdots \otimes v_{p+q} \\ &\quad + \sum_{j=p+1}^{p+q} v_1 \otimes \cdots \otimes F(v_j) \otimes \cdots \otimes v_{p+q} \\ &= D(F)(t) \otimes t' + t \otimes D(F)(t'). \end{aligned}$$

The case $p = 0$ or $q = 0$ is verified easily.

Since $T(V)$ is generated as an associative algebra by $V \cup \{1\}$, and the values of $D(F)$ on this set of generators are determined by F $(D(F)(1) = 0)$, $D(F)$ is uniquely determined on $T(V)$ by the condition that it be a derivation. □

§5. Symmetric algebras and their properties

We shall define a multiplication on the (infinite) direct sum of the spaces $S^p(V)$ of symmetric tensors of degree p $(p = 0, 1, 2, \ldots)$

$$S(V) = \bigoplus_{p=0}^{\infty} S^p(V) \quad (\text{where } S^0(V) = T^0(V) = k).$$

Then we show that, with respect to this product, $S(V)$ becomes an associative algebra and we discuss its properties.

To define a multiplication, let $t \in S^p(V)$ and $t' \in S^q(V)$. $t \otimes t' \in T^{p+q}(V)$ is not necessarily symmetric. So we apply the symmetrizer $\mathscr{S}_{p+q}$ to $t \otimes t'$. The image is called the product of t and t' and is denoted by $t \cdot t'$:

$$t \cdot t' = \mathscr{S}_{p+q}(t \otimes t').$$

EXAMPLE 2.19. (1) Let $t = v_1$ and $t' = v_2$ $(v_1, v_2 \in V)$.

$$t \cdot t' = \mathscr{S}_2(v_1 \otimes v_2) = \tfrac{1}{2}(v_1 \otimes v_2 + v_2 \otimes v_1).$$

(2) Let $t = (v_1 \otimes v_2 + v_2 \otimes v_1)/2 \in S^2(V)$ and $t' = v_3 \in S^1(V) = V$, then

$$\begin{aligned} t \cdot t' &= \mathscr{S}_3(t \otimes t') = \mathscr{S}_3(\tfrac{1}{2}(v_1 \otimes v_2 + v_2 \otimes v_1) \otimes v_3) \\ &= \tfrac{1}{6}(v_1 \otimes v_2 \otimes v_3 + v_2 \otimes v_1 \otimes v_3 + v_1 \otimes v_3 \otimes v_2 \\ &\qquad + v_3 \otimes v_1 \otimes v_2 + v_2 \otimes v_3 \otimes v_1 + v_3 \otimes v_2 \otimes v_1) \\ &= \mathscr{S}_3(v_1 \otimes v_2 \otimes v_3). \end{aligned}$$

(3) Let $t = \mathscr{S}_2(v_1 \otimes v_2)$ and $t' = \mathscr{S}_2(v_3 \otimes v_4) \in S^2(V)$, then

$$t \cdot t' = \mathscr{S}_4(t \otimes t') = \mathscr{S}_4(\mathscr{S}_2(v_1 \otimes v_2) \otimes \mathscr{S}_2(v_3 \otimes v_4))$$
$$= \mathscr{S}_4(v_1 \otimes v_2 \otimes v_3 \otimes v_4).$$

In general, for $t = \sum_p t_p$ and $t' = \sum_q t'_q$ $(t_p, t'_q \in S^p(V))$, define

$$t \cdot t' = \sum_{p,q} t_p \cdot t'_q = \sum_{k=0}^{\infty} \left(\sum_{p+q=k} \mathscr{S}_k(t_p \otimes t'_q) \right)$$
$$= \sum_{k=0}^{\infty} \mathscr{S}_k \left(\sum_{p+q=k} t_p \otimes t'_q \right).$$

The product thus defined is obviously bilinear. To prove that $S(V)$ is an associative algebra, it is necessary to show that *the product satisfies the associativity condition.* Though this could be proved directly, we shall do it by identifying $S(V)$ with a factor algebra of $T(V)$ by virtue of the property of kernel of $\mathscr{S}_p$ (§3, (c)).

First, making use of the mappings $\mathscr{S}_p$ $(p = 0, 1, 2, \ldots)$ ($\mathscr{S}_0$ = the identity mapping), we define a linear mapping $\mathscr{S}: T(V) \to S(V)$ by

$$\mathscr{S}(t) = \sum_{p=0}^{\infty} \mathscr{S}_p(t_p), \quad \text{where } T(V) \ni t = \sum t_p \ (t_p \in T^p(V)).$$

The mapping $\mathscr{S}$ is surjective and the kernel $M(V)$ of $\mathscr{S}$ is given by

$$M(V) = \left\{ t = \sum t_p \in T(V) | t_p \in M^p(V) \text{ for all } p \right\},$$

where $M^p(V) = \operatorname{Ker} \mathscr{S}_p$.

Hence $\mathscr{S}$ induces a linear isomorphism $\overline{\mathscr{S}}$ (as vector space) of the factor space $T(V)/M(V)$ onto $S(V)$. Let π be the canonical mapping of $T(V)$ into $T(V)/M(V)$. Then $\mathscr{S} = \overline{\mathscr{S}} \circ \pi$,

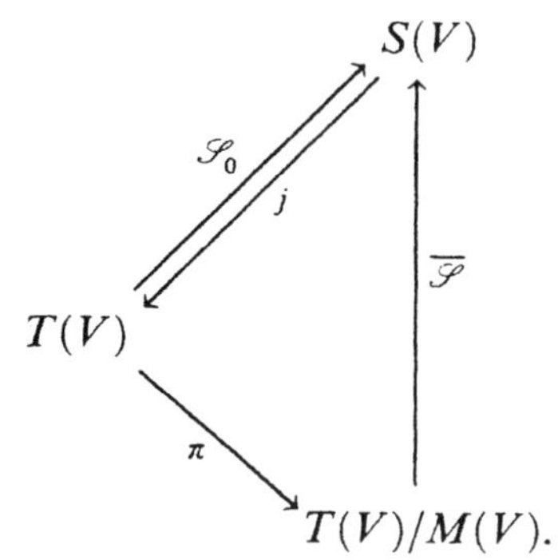

Let $\overline{\mathscr{R}}$ be the inverse mapping of $\overline{\mathscr{S}}$ and j the inclusion mapping of $S(V)$ into $T(V)$. Since each $\mathscr{S}_p$ is a projection of $T^p(V)$ onto $S^p(V)$, we can conclude that $\overline{\mathscr{R}} = \pi \circ j$. In other words, the restriction of π to $S(V)$

gives the inverse mapping $\overline{\mathcal{R}}$ of $\overline{\mathcal{S}}$. In fact, $\overline{\mathcal{S}} \circ (\pi \circ j) = \mathcal{S} \circ j =$ the identity mapping of $S(V)$. Conversely, let $x = \pi(t) \in (T(V)/M(V))$. Then $\overline{\mathcal{S}}(x) = \mathcal{S}(t)$ and $(\pi \circ j) \circ \overline{\mathcal{S}}(x) = \pi \circ j \circ \mathcal{S}(t) = \pi(\mathcal{S}(t)) = \pi(t) = x$.

The kernel $M(V)$ of $\mathcal{S}$ is, in fact, a graded ideal of the tensor algebra $T(V)$. To prove this, it suffices to verify that $M(V)$ is an ideal. Then, since $M(V) \cap T^p(V) = M^p(V)$ and $M(V)$ is a (direct) sum of $M^p(V)$, we know that it is graded. We need the following lemma.

LEMMA 2.3.

$$M^p(V) \otimes T^q(V) \subseteq M^{p+q}(V), \qquad T^q(V) \otimes M^p(V) \subseteq M^{p+q}(V).$$

PROOF. Let $v_1 \otimes \cdots \otimes v_p \in T^p(V)$, $v_{p+1} \otimes \cdots \otimes v_{p+q} \in T^q(V)$ $(v_i \in V)$ and $\tau \in \mathfrak{S}_p$. As τ operates on $\{1, 2, \dots, p\}$, we identify τ with the element $\tilde{\tau} \in \mathfrak{S}_{p+q}$ given by

$$\tilde{\tau}(\nu) = \begin{cases} \tau(\nu) & (\nu = 1, \dots, p), \\ \nu & (\nu = p+1, \dots, p+q). \end{cases}$$

Then

$$\begin{aligned} \{P_\tau(v_1 \otimes \cdots \otimes v_p)\} &\otimes (v_{p+1} \otimes \cdots \otimes v_{p+q}) \\ &= v_{\tau^{-1}(1)} \otimes \cdots \otimes v_{\tau^{-1}(p)} \otimes v_{p+1} \otimes \cdots \otimes v_{p+q} \\ &= P_{\tilde{\tau}}(v_1 \otimes \cdots \otimes v_p \otimes v_{p+1} \otimes \cdots \otimes v_{p+q}). \end{aligned}$$

By linearlity, we have, for $t \in T^p(V)$ and $t' \in T^q(V)$,

$$(P_\tau(t)) \otimes t' = P_{\tilde{\tau}}(t \otimes t').$$

$M^p(V)$ is spanned by $t - P_\tau(t)$ $(t \in T^p(V),\ \tau \in \mathfrak{S}_p)$ as a vector space (Proposition 2.8). We proved that, for $t' \in T^q(V)$

$$\begin{aligned} (t - P_\tau(t)) \otimes t' &= t \otimes t' - (P_\tau(t)) \otimes t' \\ &= t \otimes t' - P_{\tilde{\tau}}(t \otimes t'), \end{aligned}$$

which implies $(t - P_\tau(t)) \otimes t' \in M^{p+q}(V)$ since $t \otimes t' \in T^{p+q}(V)$ and $\tilde{\tau} \in \mathfrak{S}_{p+q}$. Thus we have proved the first formula. We can prove the second one by a similar argument. □

Since $M(V)$ is a graded ideal, the vector space $T(V)/M(V)$ has the structure of an associative algebra. Moreover, $T(V)/M(V)$ is a graded associative algebra. At the beginning of §5, we defined a product of two elements of $S(V)$. Since $S(V)$ can be identified with $T(V)/M(V)$ by $\overline{\mathcal{S}}$, the product of $T(V)/M(V)$ is transferred to $S(V)$. Thus we get another product on $S(V)$. But it can be shown that these two products in $S(V)$ coincide. That is, we have the following equality. For t, $t' \in S(V)$,

$$t \cdot t' = \overline{\mathcal{S}}(\overline{\mathcal{R}}(t) \cdot \overline{\mathcal{R}}(t')) \quad (\text{where } \overline{\mathcal{R}} = \overline{\mathcal{S}}^{-1}).$$

The product $\cdot$ on the right side is the multiplication in the factor algebra $T(V)/M(V)$. In fact, with π and j as above,

$$\begin{aligned} t \cdot t' &= \mathscr{S}(j(t) \otimes j(t')) && \text{(definition of product)} \\ &= \overline{\mathscr{S}}(\pi(j(t) \otimes j(t'))) && (\mathscr{S} = \overline{\mathscr{S}} \circ \pi) \\ &= \overline{\mathscr{S}}(\pi(j(t)) \cdot \pi(j(t'))) && (\pi \text{ is an (associative} \\ & && \text{algebra) homomorphism)} \\ &= \overline{\mathscr{S}}(\overline{\mathscr{R}}(t) \cdot \overline{\mathscr{R}}(t')) && (\overline{\mathscr{R}} = \pi \circ j). \end{aligned}$$

Since $T(V)/M(V)$ is an associative algebra, the product on it is associative. Therefore the product defined in $S(V)$ is also associative and we can prove that $S(V)$ is an associative algebra. Moreover, the above discussion implies that the mapping $\overline{\mathscr{S}}$ is an associative algebra isomorphism and that $\mathscr{S} = \overline{\mathscr{S}} \circ \pi \colon T(V) \to S(V)$ is an associative algebra homomorphism. $S(V) = \bigoplus_{p=0}^{\infty} S^p(V)$ is not only a graded vector space but a graded associative algebra, since we have $S^p(V) \cdot S^q(V) \subseteq S^{p+q}(V)$. The mapping $\overline{\mathscr{S}}$ assigns to a homogeneous element of $T(V)/M(V)$ a homogeneous element of $S(V)$ of the same degree.

DEFINITION 2.5. The associative algebra $S(V)$ is called the *symmetric algebra* of V.

In Proposition 2.8 we gave a set of generators of $M(V)$ as a subspace of $T(V)$. A set of generators of $M(V)$, as an ideal in $T(V)$, is given by the following proposition.

PROPOSITION 2.13. *$M(V)$ is the ideal of $T(V)$ generated by $\{v \otimes v' - v' \otimes v \mid v, v' \in V\}$.*

PROOF. Let $M'(V)$ be the ideal of $T(V)$ generated by $\{v \otimes v' - v' \otimes v \mid v, v' \in V\}$. By Proposition 2.8, the subset $\{t - P_\sigma(t)\}$ is a set of generators of $M^p(V)$ as vector space, where t ranges the set of decomposable tensors $t = v_1 \otimes \cdots \otimes v_p$ $(v_i \in V)$ and $\sigma \in X_2 = \{(1\ 2), \dots, (p-1\ p)\}$. For $\sigma = (i\ i+1)$, we have

$$\begin{aligned} t - P_\sigma(t) &= v_1 \otimes \cdots \otimes v_i \otimes v_{i+1} \otimes \cdots \otimes v_p \\ &\quad - v_1 \otimes \cdots \otimes v_{i+1} \otimes v_i \otimes \cdots \otimes v_p \\ &= v_1 \otimes \cdots \otimes v_{i-1} \otimes (v_i \otimes v_{i+1} - v_{i+1} \otimes v_i) \\ &\quad \otimes v_{i+2} \otimes \cdots \otimes v_p, \end{aligned}$$

which shows $t - P_\sigma(t) \in M'(V)$. Thus $M(V) \subseteq M'(V)$.

REMARK. If we adopt this proposition as the definition of the ideal $M(V)$, then $M(V)$ is a graded ideal by definition since it is generated by homogeneous elements. Thus, without reference to symmetric tensors we can define the symmetric algebra as the factor algebra $T(V)/M(V)$. Then, starting from this definition we can deduce what follows.

By Proposition 2.13, the following properties of $S(V)$ are easily obtained.

PROPOSITION 2.14. *Identifying $k = S^0(V)$ and $V = S^1(V)$, we can consider k and V as contained in $S(V)$. Then $S(V)$ is generated by $V \cup \{1\}$ as an associative algebra. 1 is the unit element of $S(V)$.*

Moreover, $S(V)$ is a commutative ([4]) *associative algebra.*

PROOF. Since $S(V) \cong T(V)/M(V)$ and $T(V)$ is generated by $V \cup \{1\}$, the first part is obvious.

To prove the second part, we need the following lemma.

LEMMA 2.4. *Let (A, m) be an associative algebra and S a set of generators of A. If every element of S commutes with any other element of S (i.e., $m(s, s') = m(s', s)$ for all $s, s' \in S$), A is commutative.*

PROOF. Let $s \in S$. The set of elements which commute with s is obviously a subalgebra of A. Since this subalgebra contains S, it coincides with A. Thus s is an element of the center ([5]) of A. Namely, S is contained in the center of A. Since the center is a subalgebra, the center coincides with A itself. □

For $v, v' \in V$, $v \otimes v' - v' \otimes v \in M(V)$, which implies $v \cdot v' - v' \cdot v = 0$ in $S(V)$. Thus any two elements of V ($\subseteq S(V)$) commute. Since $V \cup \{1\}$ is a set of generators of $S(V)$, it follows form Lemma 2.4 that $S(V)$ is commutative. This completes the proof of Proposition 2.14.

Using the universality of tensor algebras, we obtain the universality of symmetric algebras (for commutative associative algebras).

THEOREM 2.4. *Let A be a commutative associative algebra with the unit element e and F a linear mapping of V into A. Then F is uniquely extended to an associative algebra homomorphism $\widetilde{F}^{(S)}$ of $S(V)$ into A. That is, there exists a unique associative algebra homomorphism $\widetilde{F}^{(S)}$: $S(V) \to A$ such that $\widetilde{F}^{(S)}(1) = e$ and $\widetilde{F}^{(S)} \circ \iota_S = F$, where ι_s is the inclusion mapping of $V = S^1(V)$ into $S(V)$.*

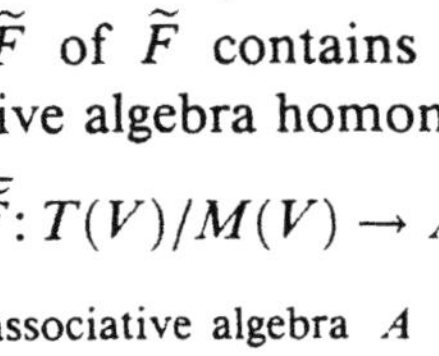

PROOF. By Theorem 2.2, F induces an associative algebra homomorphism $\widetilde{F}$ of $T(V)$ into A. If we can prove that $\widetilde{F}(t) = 0$ for all $t \in M(V)$, namely, that the kernel $\operatorname{Ker} \widetilde{F}$ of $\widetilde{F}$ contains $M(V)$, then we know that $\widetilde{f}$ naturally induces an associative algebra homomorphism

$$\widetilde{\widetilde{F}} : T(V)/M(V) \to A.$$

---

(4) If the multiplication in an associative algebra A is commutative, then A is called a commutative associative algebra.

(5) The set of elements which commutes with all elements of A is a subalgebra of A, called the center of A.

From the mapping $\overline{\mathscr{R}} = \overline{\mathscr{S}}^{-1}$, $S(V) \cong T(V)/M(V)$. Therefore $\widetilde{F}^{(S)} = \overline{\widetilde{F}} \circ \overline{\mathscr{R}}: S(V) \to A$ is an associative algebra homomorphism.

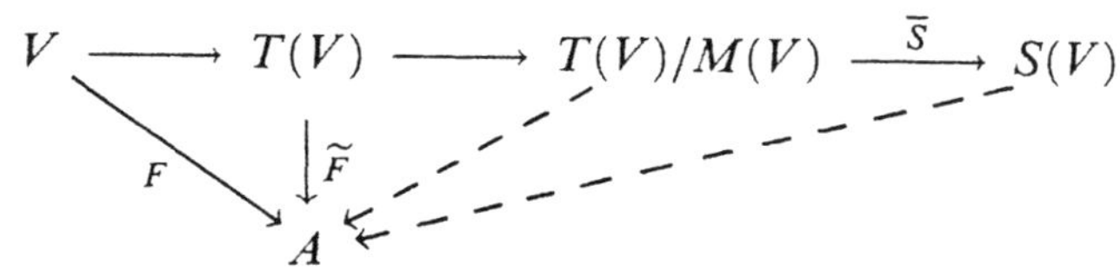

If we regard an element $t \in S(V)$ as an element of $T(V)$ (by j), $\widetilde{F}^{(S)}(t) = \widetilde{F}(t)$. In fact, $\widetilde{F}^{(S)}(t) = \overline{\widetilde{F}}(\overline{\mathscr{R}}(t)) = \overline{\widetilde{F}}(\pi \circ j(t)) = \widetilde{F}(j(t))$. In particular, for $v \in V$, we have $\widetilde{F}^{(S)}(v) = F(v)$.

Since A is commutative, for $v, v' \in V$ and $\cdot$ the product on A

$$\begin{aligned} \widetilde{F}(v \otimes v' - v' \otimes v) &= \widetilde{F}(v \otimes v') - \widetilde{F}(v' \otimes v) \\ &= F(v) \cdot F(v') - F(v') \cdot F(v) \\ &= 0. \end{aligned}$$

Thus, the set of generators of $M(V)$ given by Proposition 2.13 is contained in $\operatorname{Ker} \widetilde{F}$, which is an ideal. Therefore $M(V) \subseteq \operatorname{Ker} \widetilde{F}$, as required.

The uniqueness follows from the fact that the values of $\widetilde{F}^{(S)}$ on the set $V \cup \{1\}$ of generators are determined by F. □

NOTE. Let A, A' be associative algebras and let $\Phi: A \to A'$ be an associative algebra homomorphism. Let $\mathfrak{a} \subseteq A$ be an ideal such that $\operatorname{Ker} \Phi \supseteq \mathfrak{a}$. If we define a mapping $\overline{\Phi}: A/\mathfrak{a} \to A'$ by $\overline{\Phi}(\overline{a}) = \Phi(a)$ for $\overline{a} = a + \mathfrak{a} \in A/\mathfrak{a}$ $(a \in A)$, $\overline{\Phi}(\overline{a})$ is independent of the choice of coset representative a and $\overline{\Phi}$ is an associative algebra homomorphism.

COROLLARY. *Let V, W be k-vector spaces and $F: V \to W$ a linear mapping. Then F is uniquely extended to an associative algebra homomorphism $S(F)$ of $S(V)$ into $S(W)$.*

This assertion is similar to that of the corollary to Theorem 2.2. Let j_S be the natural inclusion mapping of W into $S(W)$. Then apply the theorem to $j_S \circ F$.

By Theorem 2.4, we can obtain a characterization of $S(V)$ by universality, namely, the analogue of Theorem 2.3 for a commutative associated algebra. The details are left to the readers.

THEOREM 2.5. *Let V be a k-vector space of dimension n. Then $S(V)$ is isomorphic to the polynomial algebra in n variables over k.*

PROOF. Take a basis $(e_1, \dots, e_n)$ of V. Let $X_1, \dots, X_n$ be n independent variables and $\mathscr{P}$ the (commutative) associative algebra consisting of polynomials in $X_1, \dots, X_n$ with coefficients in k. Let F be a linear mapping of V into $\mathscr{P}$ which assigns X_i to e_i. Then, by Theorem 2.4, F induces an associative algebra homomorphism $\widetilde{F}^{(S)}$ of $S(V)$ into $\mathscr{P}$.

For any polynomial $P = P(X_1, \dots, X_n) \in \mathscr{P}$, by construction of $\widetilde{F}^{(S)}$, we obviously have

$$\widetilde{F}^{(S)}(P(e_1, \dots, e_n)) = P(X_1, \dots, X_n) = P,$$

which implies that $\widetilde{F}^{(S)}$ is onto, and that $P(e_1, \dots, e_n) \neq 0$ for $P \neq 0$. All elements of $S(V)$ can be expressed in the form $P(e_1, \dots, e_n)$ for certain $P \in \mathscr{P}$ (cf. Proposition 2.14). Thus we have $\operatorname{Ker} \widetilde{F}^{(S)} = \{0\}$ and $\widetilde{F}^{(S)}$ is an isomorphism. □

For alternating tensors, there are results similar to those in this section. We shall describe them in Chapter 3 together with their applications.

§6. Definition of relative tensors (pseudotensor, tensor density)

In some books on tensors, the notion of tensor density or pseudotensor is discussed. In this section, after extending the concept of tensors, we give the definition of these concepts. In fact, we define a new concept which contains our tensor spaces and call it a generalized tensor space. Tensor density and pseudotensor are defined as special cases of it. Let V be a vector space. Tensor density and pseudotensor are correspondences which associate a tensor to each basis for V and which satisfy certain conditions. The conditions are given as follows: the relations between tensors associated to different bases are described by relations between corresponding bases. To explain the situation precisely, we use the notions of a set on which a group operates and of an equivariant mapping between these sets.

First, we shall define the notion of a group acting on a set. Let G be a group. The law of composition on G is denoted by $G \times G \ni (g, g') \mapsto gg' \in G$. We denote by e the unit element of G and by g^{-1} the inverse of g.

DEFINITION 2.6. Let X be a set. If there exists a mapping $X \times G \to X$, $(x, g) \mapsto x \cdot g$, which satisfies the following conditions (1) and (2), then we say that *G operates on X on the right.*

(1) For all $g, g' \in G$ and $x \in X$, $x \cdot (gg') = (x \cdot g) \cdot g'$.

(2) For all $x \in X$, $x \cdot e = x$.

On the other hand, if there exists a mapping $G \times X \to X$, $(g, x) \mapsto g \cdot x$, which satisfies the following conditions $(1)'$ and $(2)'$, then we say that G *operates on X on the left.*

$(1)'$ For all $g, g' \in G$ and $x \in X$, $(gg') \cdot x = g \cdot (g' \cdot x)$.

$(2)'$ For all $x \in X$, $e \cdot x = x$.

EXAMPLE 2.20. Let V be a k-vector space of dimension n. Let $\mathscr{F}(V)$ be the set of all bases of V. Let $\mathscr{E} = (e_1, \dots, e_n) \in \mathscr{F}(V)$ and $A = (\alpha^i{}_j)(\alpha^i{}_j \in k)$ be a regular $n \times n$ matrix. Define $e'_j \in V$ by $e'_j = \sum_{i=1}^n \alpha^i{}_j e_i$ $(j = 1, \dots, n)$. Then $\mathscr{E}' = (e'_1, \dots, e'_n)$ is also an element of $\mathscr{F}(V)$. If we define $\mathscr{E}' = \mathscr{E} \cdot A$, the group $\mathrm{GL}_n(k)$ of regular $n \times n$ matrices over k operates on $\mathscr{F}(V)$ on the right.

In this case, we can easily observe that, for all $\mathscr{E}, \mathscr{E}' \in \mathscr{F}(V)$, there always exists a unique $A \in \mathrm{GL}_n(k)$ such that $\mathscr{E}' = \mathscr{E} \cdot A$. In particular, from $\mathscr{E} \cdot A = \mathscr{E}$, it necessarily follows that $A = I$, the identity matrix.

EXAMPLE 2.21. Let U be a k-vector space and $\mathrm{GL}(U)$ the group of regular linear transformations of U. For $u \in U$ and $g \in \mathrm{GL}(U)$, define $g \cdot u = g(u)$. Then $\mathrm{GL}(U)$ operates on U on the left. More generally, let G be a group and $\rho: G \to \mathrm{GL}(U)$ a group homomorphism (6), namely $\rho(gg') = \rho(g)\rho(g')$ for all $g, g' \in G$. Define $g \cdot u = \rho(g)(u)$ for $g \in G$ and $u \in U$. Then G operates on U on the left.

EXAMPLE 2.22. Let G be a group and X a set on which G operates on the left. If we define a mapping $X \times G \to X$ by $(x, g) \mapsto g^{-1} \cdot x$, then G operates on X on the right by this mapping.

Next, we define equivariant mappings as follows.

DEFINITION 2.7. Let X and Y be sets on which a group G operates on the right. A mapping f of X into Y is called *G-equivariant*, if, for all $x \in X$ and $g \in G$, we have

$$f(x \cdot g) = f(x) \cdot g.$$

(Note that $\cdot g$ on the left denotes the operation on X, and $\cdot g$ on the right denotes that on Y.)

To extend the notion of tensor space, we consider the following set and operation of a group.

DEFINITION 2.8 (GT). Let V be a k-vector space of dimension n, U a k-vector space, and ρ a group homomorphism of $G = \mathrm{GL}_n(k)$ into $\mathrm{GL}(U)$. Then, by Example 2.20, G operates on the set $\mathscr{F}(U)$ of bases of V on the right. Moreover, G operates on U on the right by

$$u \cdot A = \rho(A^{-1})(u) \qquad (u \in U,\ A \in G)$$

(See Examples 2.21 and 2.22.) We denote by $\mathscr{T}(V, U)$ the set of G-equivariant mappings of $\mathscr{F}(V)$ into U. Namely, an element f of $\mathscr{T}(V, U)$ assigns an element of U to every basis of V, satisfying the condition of G-equivariance.

We have the next lemma with respect to $\mathscr{T}(V, U)$.

LEMMA 2.5. *Let us fix an element $\mathscr{E}_0 \in \mathscr{F}(V)$. Then the mapping $\tilde{\mathscr{E}}_0$: $\mathscr{T}(V, U) \to U$ which assigns $f(\mathscr{E}_0) \in U$ to $f \in \mathscr{T}(V, U)$ is onto and one-to-one.*

PROOF. Let $f \in \mathscr{T}(V, U)$. For any $\mathscr{E} \in \mathscr{F}(V)$, there exists $A \in G = \mathrm{GL}_n(k)$ such that $\mathscr{E} = \mathscr{E}_0 \cdot A$. Since f is G-equivariant, we have $f(\mathscr{E}) = f(\mathscr{E}_0 \cdot A) = f(\mathscr{E}_0) \cdot A$, which means that the mapping f is determined by $f(\mathscr{E}_0)$. Therefore $\tilde{\mathscr{E}}_0$ is one-to-one.

Next, we shall show that $\tilde{\mathscr{E}}_0$ is an onto mapping. Let $u \in U$. We shall construct $f \in \mathscr{T}(V, U)$ such that $f(\mathscr{E}_0) = u$. For $\mathscr{E} \in \mathscr{F}(V)$, there exists a unique $A \in G$ such that $\mathscr{E} = \mathscr{E}_0 \cdot A$. Using A, define $f(\mathscr{E}) = u \cdot A$. In particular, $f(\mathscr{E}_0) = u \cdot I = u$.

(6) P is called a representation of G on U.

To show that f is G-equivariant, let $\mathscr{E} \in \mathscr{F}(V)$ and $A_1 \in G$. Set $\mathscr{E} \cdot A_1 = \mathscr{E}_1$. There exists A, $A' \in G$ such that $\mathscr{E} = \mathscr{E}_0 \cdot A$, and $\mathscr{E}_1 = \mathscr{E}_0 \cdot A'$. Then, by definition, $f(\mathscr{E}) = u \cdot A$ and $f(\mathscr{E}_1) = u \cdot A'$. On the other hand, since $\mathscr{E}_1 = \mathscr{E} \cdot A_1 = (\mathscr{E}_0 \cdot A) \cdot A_1 = \mathscr{E}_0 \cdot (A\, A_1)$ and $\mathscr{E}_1 = \mathscr{E}_0 \cdot A'$, we have $A' = A\, A_1$. Thus

$$f(\mathscr{E} \cdot A_1) = f(\mathscr{E}_1) = u \cdot A',$$
$$f(\mathscr{E}) \cdot A_1 = (u \cdot A) \cdot A_1 = u \cdot (A\, A_1) = u \cdot A',$$

which implies that f is G-equivariant. □

By the mapping $\widetilde{\mathscr{E}_0}$ defined in Lemma 2.5, we can identify $\mathscr{T}(V, U)$ with U and transfer the addition and scalar multiplication of U to $\mathscr{T}(V, U)$. Thus $\mathscr{T}(V, U)$ becomes a k-vector space. In other words, for $f_1, f_2 \in \mathscr{T}(V, U)$ and $\alpha \in k$, $f_1 + f_2$ and αf_1 are defined to be the elements in $\mathscr{T}(V, U)$ whose values for $\mathscr{E}_0$ are $f_1(\mathscr{E}_0) + f_2(\mathscr{E}_0)$, $\alpha(f_1(\mathscr{E}_0))$ respectively. Therefore the mapping $\widetilde{\mathscr{E}_0}$ is a linear isomorphism.

LEMMA 2.6. *The sum and scalar multiple defined above do not depend on the choice of* $\mathscr{E}_0$.

PROOF. Let $\mathscr{E}_0' \in \mathscr{F}(V)$, and let $A \in G$ such that $\mathscr{E}_0 \cdot A = \mathscr{E}_0'$. For f_1 and $f_2 \in \mathscr{T}(V, U)$, the following computation shows that the value of $f_1 + f_2$ defined by $\mathscr{E}_0$ for $\mathscr{E}_0'$ is equal to $f_1(\mathscr{E}_0') + f_2(\mathscr{E}_0')$, which implies that $f_1 + f_2$ defined by $\mathscr{E}_0$ coincides with that defined by $\mathscr{E}_0'$.

$$\begin{aligned}
(f_1 + f_2)(\mathscr{E}_0') &= (f_1 + f_2)(\mathscr{E}_0 \cdot A) = ((f_1 + f_2)(\mathscr{E}_0)) \cdot A \\
&= (f_1(\mathscr{E}_0) + f_2(\mathscr{E}_0)) \cdot A \\
&= \rho(A^{-1})(f_1(\mathscr{E}_0) + f_2(\mathscr{E}_0)) \\
&= \rho(A^{-1})(f_1(\mathscr{E}_0)) + \rho(A^{-1})(f_2(\mathscr{E}_0)) \\
&= f_1(\mathscr{E}_0) \cdot A + f_2(\mathscr{E}_0) \cdot A \\
&= f_1(\mathscr{E}_0 \cdot A) + f_2(\mathscr{E}_0 \cdot A) \\
&= f_1(\mathscr{E}_0') + f_2(\mathscr{E}_0').
\end{aligned}$$

The proof is similar to this in the case of scalar multiplication. □

Thus we can regard $\mathscr{T}(V, U)$ as k-vector space. By definition, to obtain certain property of $\mathscr{T}(V, U)$ as k-vector space, we may fix an $\mathscr{E}_0$ and transfer the corresponding one of U. For instance, let $(g_1, \ldots, g_m)$ be a basis for U. Set $f_i = \widetilde{\mathscr{E}_0}^{-1}(g_i)$ (i.e., an element $f_i \in \mathscr{T}(V, U)$ such that $f_i(\mathscr{E}_0) = g_i$). Then $(f_1, \ldots, f_m)$ is a basis for $\mathscr{T}(V, U)$.

EXAMPLE 2.23. Define $\rho(A) =$ the identity transformation of U for all A of $G = \mathrm{GL}_n(k)$. Then $u \cdot A = \rho(A^{-1})u = u$ for all $u \in U$. Let $f \in \mathscr{T}(V, U)$. We have $f(\mathscr{E}_1) = f(\mathscr{E}_0 \cdot A) = f(\mathscr{E}_0) \cdot A = f(\mathscr{E}_0)$ for all $\mathscr{E}_1 \in \mathscr{F}(V)$, which means that f assigns the same value to every $\mathscr{E} \in \mathscr{F}(V)$. Identifying f with this value, we may regard $\mathscr{T}(V, U) = U$. In particular, $\mathscr{T}(V, U) = T_q{}^p(V)$ for $U = T_q{}^p(V)$.

Using these facts, we shall define the notion of generalized tensor space as follows.

DEFINITION 2.9. Let V, U, ρ, $\mathcal{F}(V)$, the operation of $G = \mathrm{GL}_n(k)$ on U, and $\mathcal{T}(V, U)$ be defined as in Definition 2.8. If U is a tensor space $T_q^{\,p}(V)$, then the vector space $\mathcal{T}(V, U)$ is called the *generalized tensor space of type* (U, ρ) and an element of $\mathcal{T}(V, U)$ is called a generalized tensor of type (U, ρ).

DEFINITION 2.10. Let $U = T_q^{\,p}(V)$ and let $f: \mathcal{F}(V) \to U$ be a generalized tensor of type (U, ρ).For $\mathscr{E} \in \mathcal{F}(V)$, the components of the tensor $f(\mathscr{E})$ of type (p, q) with respect to $\mathscr{E}$ are called the *components* of f with respect to $\mathscr{E}$.

EXAMPLE 2.24. Let t be an integer and $U = T_q^{\,p}(V)$. Define $\rho(A) = (\det(A))^t \times$ (the identity transformation of U) for $A \in \mathrm{GL}_n(k)$. By a property of the determinant, ρ is a homomorphism of $\mathrm{GL}_n(k)$ into $\mathrm{GL}(U)$. Let $\mathscr{E}_1, \mathscr{E}_2 \in \mathcal{F}(V)$ and write $\mathscr{E}_2 = \mathscr{E}_1 \cdot A$ $(A \in G)$. Then, for $f \in \mathcal{T}(V, U)$, we have $f(\mathscr{E}_2) = f(\mathscr{E}_1) \cdot A = \rho(A^{-1}) f(\mathscr{E}_1) = (\det(A))^{-t} f(\mathscr{E}_1)$.

We shall compute relations between the components of a generalized tensor of type (U, ρ). Let $\mathscr{E}_1, \mathscr{E}_2 \in \mathcal{F}(V)$ and write $\mathscr{E}_2 = \mathscr{E}_1 \cdot A$, where $A = (\alpha^i{}_j)$. Let $f \in \mathcal{T}(V, U)$ and $\{\xi_{j_1, \dots, j_q}{}^{i_1, \dots, i_p}\}$ be the components of f with respect to $\mathscr{E}_1$, which are the components of the tensor $f(\mathscr{E}_1)$ with respect to $\mathscr{E}_1$. By definition $f(\mathscr{E}_2) = (\det(A))^{-t} f(\mathscr{E}_1)$. Since the components of $f(\mathscr{E}_1)$ with respect to $\mathscr{E}_2$ are

$$\left\{\sum \alpha^{l_1}{}_{j_1} \cdots \alpha^{l_q}{}_{j_q} \beta^{i_1}{}_{k_1} \cdots \beta^{i_p}{}_{k_p} \xi_{l_1, \dots, l_q}{}^{k_1, \dots, k_p}\right\},$$

where $(\beta^i{}_j) = (\alpha^i{}_j)^{-1}$ (see (2.4)), the components $\{\xi'_{j_1, \dots, j_q}{}^{i_1, \dots, i_p}\}$ of f with respect to $\mathscr{E}_2$ (= the components of $f(\mathscr{E}_2)$ with respect to $\mathscr{E}_2$) are given by

$$\xi'_{j_1, \dots, j_q}{}^{i_1, \dots, i_p} = (\det(\alpha^i{}_j))^{-t} \sum \alpha^{l_1}{}_{j_1} \cdots \alpha^{l_q}{}_{j_q} \beta^{i_1}{}_{k_1} \cdots \beta^{i_p}{}_{k_p} \xi_{l_1, \dots, l_q}{}^{k_1, \dots, k_p}.$$

The generalized tensors of this type are called the *relative tensors of type* (p, q) *and weight* $(-t)$ in some of the literature.

In a similar fashion, let $k = \mathbf{R}$, the field of real numbers. Instead of the above homomorphism ρ, consider the following homomorphisms ρ_1, ρ_2, and ρ_3.

$$\rho_1(A) = |\det(A)|^t \times (\text{the identity transformation of } U),$$

$$\rho_2(A) = (\text{the sign of } \det(A)) \times |\det(A)|^t \times (\text{the identity transformation of } U),$$

$$\rho_3(A) = (\text{the sign of } \det(A)) \times (\text{the identity transformations of } U),$$

(where t is an integer and the sign of $\det(A)$ is 1 or -1 according to $\det(A) > 0$ or $\det(A) < 0$.)

Then the corresponding generalized tensors are called *tensor densities of type* (p, q) *and weight* $(-t)$ in the cases of ρ_1 or ρ_2 and *pseudotensors* ([7]) *of type* (p, q) in the case of ρ_3.

EXAMPLE 2.25. Let Φ be a bilinear form on $V \times V$ and $U = k$. For $\mathscr{E} = (e_1, \dots, e_n) \in \mathscr{F}(V)$, define $f(\mathscr{E}) = \det(\Phi(e_i, e_j))$. f is a relative scalar (= relative tensor of type $(0, 0)$) of weight 2. In fact, for $\mathscr{E}' = \mathscr{E} \cdot A = (e'_1, \dots, e'_n)$, we have

$$(\Phi(e'_i, e'_j)) = {}^tA(\Phi(e_i, e_j))A,$$

and $f(\mathscr{E}') = (\det(A))^2 f(\mathscr{E})$.

EXAMPLE 2.26. The operation of the *vector product* (or cross product) in Euclidean space of dimension 3 can be regarded as pseudovector ([8]) (= pseudotensor of type $(1, 0)$). First, we shall recall the definition of vector product. Let V be Euclidean space of dimension 3, namely, 3-dimensional real vector space with a positive definite inner product $(\ ,\ )$. Define the length $\|v\|$ of a vector $v \in V$ by $\|v\| = \sqrt{(v, v)}$. Fix a bases $\mathscr{E}_0$ of V. For $v_1, v_2 \in V$, define $v_3 = v_1 \times v_2 \in V$ by the following conditions: if $\{v_1, v_2\}$ is linearly dependent, $v_3 = 0$. If not, the length $\|v_3\|$ is given by $\|v_3\| = \|v_1\| \cdot \|v_2\| \sin\theta$ (where θ is the angle between v_1 and v_2), $(v_1, v_3) = (v_2, v_3) = 0$, and the determinant of the matrix of the change of basis from $\mathscr{E}_0$ to (v_1, v_2, v_3) is positive. The element $v_1 \times v_2$ of V depends on the choice of $\mathscr{E}_0$. Set $w = v_1 \times v_2$ for $\mathscr{E}_0$, then for another basis $\mathscr{E}_1$, $v_1 \times v_2$ (with respect to $\mathscr{E}_1$) is given by $v_1 \times v_2 =$ (the sign of $(\det(A)) \times w$, where A is the matrix of the change of basis from $\mathscr{E}_0$ to $\mathscr{E}_1$.

Let V be as above and $U = V$. Define $\rho_1 : GL_3(\mathbf{R}) \to GL(V)$ by $\rho_1(A) =$ (the sign of $\det(A)$)$\times$ the identity transformation of V). Let f be an element of $\mathscr{T}(V, V)$ such that $f(\mathscr{E}_0) = w$. Then $f(\mathscr{E}_1) = f(\mathscr{E}_0 \cdot A) =$ (the sign of $\det(A)$) $\times\, w$. Thus we can consider that the vector product $(v_1, v_2) \mapsto v_1 \times v_2$ is a bilinear mapping $V \times V \to \mathscr{T}(V, V)$.

EXAMPLE 2.27. Let $U = T^n(V)$. For $\mathscr{E} = (e_1, \dots, e_n) \in \mathscr{F}(V)$, define

$$f(\mathscr{E}) = n!\mathscr{A}_n(e_1 \otimes \cdots \otimes e_n),$$

then we have a mapping $f : \mathscr{F}(V) \to U$. If $\mathscr{E}' = \mathscr{E} \cdot A$ $(A \in GL_n(k))$, $f(\mathscr{E}') = \det(A) \cdot f(\mathscr{E})$. (Though this can be proved by direct computation, it is easily obtained by Lemma 3.1 and Proposition 3.3.) Therefore f is a relative tensor of type $(n, 0)$ and weight 1.

Let us compute the components of this tensor. By definition, for any basis

---

([7]) Some authors use differently the terms relative tensor, tensor density, and pseudotensor.

([8]) A pseudovector is sometimes called an axial vector.

$\mathscr{E} = (e_1, \dots, e_n)$,

$$\begin{aligned} f(\mathscr{E}) &= n!\mathscr{A}_n(e_1 \otimes \cdots \otimes e_n) \\ &= \sum_{\sigma \in \mathfrak{S}_n} \operatorname{sgn}(\sigma)(e_{\sigma(1)} \otimes \cdots \otimes e_{\sigma(n)}). \end{aligned}$$

Thus, the components $\{\xi^{i_1, \dots, i_n}\}$ are given by

$$\xi^{i_1, \dots, i_n} = \begin{cases} \operatorname{sgn}(\binom{1 \cdots n}{i_1 \cdots i_n}) & (\text{if all } i_j\text{s are different}), \\ 0 & (\text{otherwise}). \end{cases}$$

This relative tensor is sometimes called *Eddington's epsilon.*

Exercises

In Exercises 1 to 4, let $V = \mathbf{R}^3$ be the real vector space consisting of all the column vectors of dimension 3. Take the basis $\mathscr{E} = (\mathbf{e}_1, \mathbf{e}_2, \mathbf{e}_3)$ of V consisting of the unit vectors. Elements of V^* will be expressed by row vectors using the correspondence $\varphi \to (\varphi(\mathbf{e}_1), \varphi(\mathbf{e}_2), \varphi(\mathbf{e}_3))$.

1. Let $\mathscr{E}' = (\mathbf{e}_1', \mathbf{e}_2', \mathbf{e}_3')$ be a basis of V defined by

$$\begin{cases} \mathbf{e}_1' = \mathbf{e}_1 - \mathbf{e}_2 + 2\mathbf{e}_3, \\ \mathbf{e}_2' = 2\mathbf{e}_1 + \mathbf{e}_2 - \mathbf{e}_3, \\ \mathbf{e}_3' = \mathbf{e}_1 + \mathbf{e}_2 + \mathbf{e}_3. \end{cases}$$

(1) Let t_1 be a contravariant tensor of degree 2, whose components $\{\xi^{ij}\}$ $(1 \le i, j \le 3)$ with respect to $\mathscr{E}$ are given by

$$(\xi^{ij}) = \begin{bmatrix} 1 & 2 & 3 \\ 2 & -1 & -3 \\ -3 & 1 & 4 \end{bmatrix}.$$

Compute the components of t_1 with respect to $\mathscr{E}'$.

(2) Let t_2 be a tensor of type $(1, 1)$, whose components $\{\xi_i^{\ j}\}$ with respect to $\mathscr{E}$ are given by

$$(\xi_i^{\ j}) = \begin{bmatrix} 1 & 1 & 0 \\ 2 & 1 & 2 \\ 3 & 2 & 1 \end{bmatrix}.$$

Compute the components of t_2 with respect to $\mathscr{E}'$.

2. Recall the isomorphism $V^* \otimes V \cong \operatorname{Hom}(V, V)$ given in Example 2.1 §1.

(1) Let T_1 be the linear transformation of V corresponding to the following element of $V^* \otimes V$,

$$(-1, 2, 3) \otimes \begin{bmatrix} 2 \\ 1 \\ 4 \end{bmatrix} + (2, 2, 1) \otimes \begin{bmatrix} 1 \\ 2 \\ 1 \end{bmatrix}.$$

Give the matrix of T_1 with respect to $\mathscr{E}$.

(2) Let T_2 be the element of $\mathrm{Hom}(V, V)$, whose matrix with respect to $\mathscr{E}$ is given by

$$\begin{bmatrix} 1 & 1 & -1 \\ 2 & 1 & 4 \\ 3 & 2 & 1 \end{bmatrix}.$$

Compute the element of $V^* \otimes V$ corresponding to T_2 by the above isomorphism.

(3) When you write the element obtained in (2) as a sum of $\varphi_i \otimes v_i$ $(\varphi_i \in V^*, \ v_i \in V)$, try to minimize the number of terms.

3. (1) Let $A = (\alpha_{ij})$ be a 3×3 matrix with coefficients in $\mathbf{R}$. The mapping $\Phi_A: V \times V \to \mathbf{R}$ defined by $\Phi_A(v, w) = {}^t vAw$ is an element of $\mathscr{L}(V, V; \mathbf{R})$. Using α_{ij}, write in the form of a tensor product of row vectors, the element which corresponds to Φ_A by the isomorphism $\mathscr{L}(V, V; \mathbf{R}) \cong V^* \otimes V^*$.

(2) If $A = (\alpha_{ij})$ is a symmetric matrix, then Φ_A is a symmetric bilinear form, to which corresponds a symmetric tensor of $V^* \otimes V^*$. Compute this symmetric tensor.

4. Let $t_1 \in T^2(V)$ be the tensor given in (1), exercise 1. Compute the components of the tensors $\mathscr{S}(t_1)$ and $\mathscr{A}(t_1)$ with respect to $\mathscr{E}$.

In the following exercises, let V be a k-vector space.

5. By Proposition 2.6, $\{\mathscr{S}_p(t(i_1, \dots, i_p)) | i_1 \leq \cdots \leq i_p\}$ form a basis of $S^p(V)$. We can express $t \in S^p(V)$ as linear combination of basis elements

$$t = \sum_{i_1 \leq \cdots \leq i_p} \alpha^{i_1, \dots, i_p}(\mathscr{S}_p(t(i_1, \dots, i_p))).$$

On the other hand, since $t \in S^p(V) \subseteq T^p(V)$, we have

$$t = \sum_{j_1, \dots, j_p} \beta^{j_1, \dots, j_p} t(j_1, \dots, j_p)$$

using the standard basis. Analyze the relations between $\{\alpha^{i_1, \dots, i_p}\}$ and $\{\beta^{j_1, \dots, j_p}\}$.

Discuss a similar problem for an element of $A^p(V)$ using Proposition 2.7.

6. Let $G: V \times V \to k$ be a nondegenerate bilinear form. (That is, we have $\det(G(e_i, e_j)) \neq 0$ for all bases $(e_1, \dots, e_n)$ of V.)

(1) For $v_0 \in V$, define $\varphi_{v_0}(v) = G(v, v_0)$ $(v \in V)$. Then, show that $\varphi_{v_0} \in V^*$ and that the mapping $\widetilde{G}: v_0 \mapsto \varphi_{v_0} \in V^*$ is a linear isomorphism of V onto V^*.

(2) For $F \in \mathscr{L}(\underbrace{V, \dots, V}_{q \text{ terms}}, \underbrace{V^*, \dots, V^*}_{p \text{ terms}}; k)$, define

$$\overline{F} \in \mathscr{L}(\underbrace{V, \dots, V}_{(q-1) \text{ terms}}, \underbrace{V^*, \dots, V^*}_{(p+1) \text{ terms}}; k)$$

and

$$\underline{F} \in \mathscr{L}(\underbrace{V, \dots, V}_{(q+1)\text{ terms}}, \underbrace{V^*, \dots, V^*}_{(p-1)\text{ terms}}; k)$$

by

$$\begin{aligned}\overline{F}(v_1, \dots, v_{q-1}, \varphi_1, \dots, \varphi_{p+1}) \\ &= F(v_1, \dots, v_{q-1}, \widetilde{G}^{-1}(\varphi_1), \varphi_2, \dots, \varphi_{p+1}), \\ \underline{F}(v_1, \dots, v_q, v_{q+1}, \varphi_1, \dots, \varphi_{p-1}) \\ &= F(v_1, \dots, v_q, \widetilde{G}(v_{q+1}), \varphi_1, \dots, \varphi_{p-1}).\end{aligned}$$

Let $\{\xi_{j_1, \dots, j_q}{}^{i_1, \dots, i_p}\}$ be the components of F with respect to a basis $\mathscr{E} = (e_1, \dots, e_n)$ of V, where F is regarded as element of $T_q{}^p(V)$. Compute the components of $\overline{F} \in T_{q-1}{}^{p+1}(V)$ and $\underline{F} \in T_{q+1}{}^{p-1}(V)$, where we set $G(e_i, e_j) = g_{ij}$ and ${}^t(g_{ij})^{-1} = (g^{ij})$.([9])

([9]) When such a G is given, the tensors F, $\underline{F}$, and $\overline{F}$ can be considered as different forms of the same object. These operations are called operations of *raising and lowering indices* (using G).

CHAPTER III

Exterior Algebra and its Applications

In the previous chapter, we defined the tensor algebra of a vector space V and constructed the symmetric algebra, giving a product in the space of symmetric tensors. By applying a similar argument to the space of alternating tensors, we can construct the exterior algebra (Grassmann algebra).

The concept of exterior algebra is fundamental to the theory of differential forms used in geometry and analysis. In §1, we give a thorough discussion of the definition, though some parts may repeat some of what we have already discussed. In §2, we present applications to determinant. In §4, we give applications to geometry and discuss the correspondence between subspaces of V and elements of exterior algebras.

§1. Definition of exterior algebra and its properties

We denote by $A^p(V)$ the space of all alternating tensors of degree p. Recall that $A^1(V) = T^1(V) = V$ and $A^m(V) = \{0\}$ for $m > n = \dim V$ (see §2.3(b)). Let $A(V)$ be the direct sum of $A^p(V)$, where $A^0(V) = T^0(V) = k$.

$$A(V) = \bigoplus_{p=0}^{\infty} A^p(V) = \bigoplus_{p=0}^{n} A^p(V).$$

Next, we define a multiplication on $A(V)$. Let $t \in A^p(V)$ and $t' \in A^q(V)$. Then, $t \otimes t'$ is an element of $T^{p+q}(V)$ and we can apply the alternator $\mathscr{A}_{p+q}$. The element $\mathscr{A}_{p+q}(t \otimes t')$ is called the *exterior product* of t and t' and is denoted by $t \wedge t'$. If $p+q > n$, the alternator $\mathscr{A}_{p+q}$ is the zero mapping. This product is extended to all elements of $A(V)$ by bilinearlity. Namely, for $t = \sum t_p$, $t' = \sum t'_p$ $(t_p, t'_p \in A^p(V))$ of $A(V)$, $t \wedge t'$ is defined by

$$t \wedge t' = \sum_{p,q} (t_p \wedge t_q) = \sum_{k=0}^{n} \mathscr{A}_k \left(\sum_{p+q=k} t_p \otimes t'_q \right).$$

EXAMPLE 3.1. Let v, $v' \in V = A^1(V)$. Then,

$$v \wedge v' = \mathscr{A}_2(v \otimes v') = \frac{1}{2}(v \otimes v' - v' \otimes v).$$

We also have,

$$v' \wedge v = \mathscr{A}_2(v' \otimes v) = \mathscr{A}_2 P_\sigma(v \otimes v') = \operatorname{sgn}(\sigma)\mathscr{A}_2(v \otimes v')$$
$$= -(v \wedge v')$$

(where σ is the transposition (1.2)). In particular, the exterior product on $A(V)$ is not commutative. Setting $v' = v$, we have $v \wedge v = 0$.

To verify that the exterior product $\wedge$ satisfies associativity, we show that, as in the case of a symmetric algebra, $A(V)$ can be identified with a factor algebra of $T(V)$. Though we could proceed in exactly the same way as with symmetric algebras, we present an alternate approach.

From Example 3.1, we have that $v \wedge v = 0$ for $v \in A^1(V) = V$. Consider the ideal generated by $S = \{v \otimes v | v \in V\}$ of tensor algebra $T(V) = \bigoplus_{p=0}^{\infty} T^p(V)$. Denote this ideal by $I(V)$. Since all elements of S are homogeneous, $I(V)$ is by Proposition 2.11 a graded ideal. Set $I^p(V) = I(V) \cap T^p(V)$ Then $I(V) = \bigoplus_{p=0}^{\infty} I^p(V)$ and the factor algebra $\mathfrak{A}(V) = T(V)/I(V)$ is a graded algebra $\mathfrak{A}(V) \cong \bigoplus_{p=0}^{\infty}(T^p(V)/I^p(V))$. Let π be the natural homomorphism $T(V) \to \mathfrak{A}(V)$ which assigns to $t \in T(V)$ the coset $t+I(V) \in \mathfrak{A}(V)$. Since $S \subseteqq T^2(V)$, we obviously have $I^0(V) = I^1(V) = \{0\}$. Thus the restriction of π to V: $V = T^1(V) \to \mathfrak{A}(V)$ is a one-to-one mapping.

To simplify the notation, we write $\pi(t) = \bar{t}$ ($t \in T(V)$ and denote by $\bar{\wedge}$ the product in the factor algebra $\mathfrak{A}(V)$. Namely, for t_1, $t_2 \in T(V)$, $\bar{t}_1 \wedge \bar{t}_2 = \pi(t_1 \otimes t_2)$. For v_1, $v_2 \in V$,

$$v_1 \otimes v_2 + v_2 \otimes v_1$$
$$= (v_1 + v_2) \otimes (v_1 + v_2) - (v_1 \otimes v_1) - (v_2 \otimes v_2) \in I(V).$$

Thus

$$\bar{v}_1 \,\bar{\wedge}\, \bar{v}_2 + \bar{v}_2 \,\bar{\wedge}\, \bar{v}_1 = 0$$

i.e., $\bar{v}_1 \,\bar{\wedge}\, \bar{v}_2 = -\bar{v}_2 \,\bar{\wedge}\, \bar{v}_1$. By definition, we also have

$$\bar{v}_1 \,\bar{\wedge}\, \bar{v}_1 = \bar{0}.$$

Let $v_1 \otimes \cdots \otimes v_p \in T^p(V)$. If there are indices i and j ($i < j$) such that $v_i = v_j$, then we have([1])

$$\pi(v_1 \otimes \cdots \otimes v_p) = \bar{v}_1 \,\bar{\wedge} \cdots \bar{\wedge}\, \bar{v}_p = \bar{0},$$

by repeated use of the above formulas. This means that $v_1 \otimes \cdots \otimes v_p \in I^p(V)$. In fact, we have

PROPOSITION 3.1. *Let N'_p be the vector subspace of $T^p(V)$ spanned by all elements $v_1 \otimes \cdots \otimes v_p$ ($v_i \in V$) such that $v_i = v_j$ for some $i \neq j$. Then N'_p is equal to $I^p(V)$.*

([1])Since the product $\bar{\wedge}$ is associative, we can define the product $\bar{v}_1 \bar{\wedge} \cdots \bar{\wedge} \bar{v}_p$ of p elements.

PROOF. We have already shown that $N_p' \subseteqq I^p(V)$. Each element of $I(V)$ has the form $\sum_{i=1}^k t_i \otimes v_i \otimes v_i \otimes t_i'$, $(t_i, t_i' \in T(V))$. If necessary, express t_i, t_i' as sums of elements of $T^q(V)$. Then we know that $I^p(V) \subseteqq N_p'$. □

By Proposition 2.9, $N_p' = \operatorname{Ker}\mathscr{A}_p$. Thus $I^p(V) = \operatorname{Ker}\mathscr{A}_p$. Since $\mathscr{A}_p$ is a projection of $T^p(V)$ onto the subspace $A^p(V)$ (cf. Proposition 2.5), we have that $T^p(V) = A^p(V) \oplus I^p(V)$ (direct sum) and the restriction $\pi|A^p(V)$ of π to $A^p(V)$ gives a linear isomorphism of $A^p(V)$ onto $T^p(V)/I^p(V)$. In particular, $\pi|A^p(V)$ is one-to-one and, for all $t \in T^p(V)$, we have $\pi(t) = \pi(\mathscr{A}_p(t))$. Therefore, the restriction of π to $A(V)$ gives a linear isomorphism $\mathscr{R}: A(V) \to \mathfrak{A}(V)$.

Now we know that the exterior product of $A(V)$ defined at the beginning of this section is nothing but the product obtained by transferring the product in $\mathfrak{A}(V)$ by $\mathscr{R}$. That is, we have, for t, $t' \in A(V)$,

$$t \wedge t' = \mathscr{R}^{-1}(\mathscr{R}(t) \barwedge \mathscr{R}(t')).$$

By the linearlity of $\mathscr{R}$, it suffices to prove the equality above for $t \in A^p(V)$ and $t' \in A^q(V)$, which is done by the following equalities.

$$\begin{aligned} \mathscr{R}(t) \barwedge \mathscr{R}(t') &= \pi(t) \barwedge \pi(t') = \pi(t \otimes t') \\ &= \pi(\mathscr{A}_{p+q}(t \otimes t')) = \mathscr{R}(t \wedge t'). \end{aligned}$$

Thus $\mathscr{R}$ preserves the product. In particular, we know that $A(V)$ can be regarded as graded associative algebra $(A(V) = \bigoplus_{p=0}^\infty A^p(V))$ with respect to the exterior product and that $\mathscr{R}$ is an isomorphism of associative algebras.

By summing up the mappings $\mathscr{A}_p$ $(p = 0, 1, \dots)$, where $\mathscr{A}_0$ is the identity mapping, we define $\mathscr{A}: T(V) \to A(V)$ as follows: For $T(V) \ni t = \sum_p t_p$ $(t_p \in T^p(V))$,

$$\mathscr{A}(t) = \sum_{p=0}^\infty \mathscr{A}_p(t_p).$$

Since $\pi(t) = \pi(\mathscr{A}(t))$, we have $\mathscr{R}^{-1}(\pi(t)) = \mathscr{A}(t)$, i.e., $\mathscr{A} = \mathscr{R}^{-1} \circ \pi$, which shows that $\mathscr{A}$ is a homomorphism of associative algebras.

DEFINITION 3.1. The associative algebra $A(V)$ is called the *exterior algebra* or the *Grassmann algebra* of V.

Since the mapping $\mathscr{A}$ is a homomorphism of associative algebras, we have the following lemma.

LEMMA 3.1. *Let* $v_i \in V$ $(i = 1, \dots, p)$.

$$\mathscr{A}(v_1 \otimes \cdots \otimes v_p) = v_1 \wedge \cdots \wedge v_p.$$

In particular, for a standard basis element $t(i_1, \dots, i_p)$ *of* $T^p(V)$ *corresponding to a basis* $\mathscr{E} = (e_i)$ *of* V,

$$\mathscr{A}(t(i_1, \dots, i_p)) = e_{i_1} \wedge \cdots \wedge e_{i_p}.$$

PROOF. It is obvious by the fact that $\mathscr{A}(v) = v$ for $v \in V$. □

EXAMPLE 3.2. Since the exterior product is associative, the right sides of the equalities of Lemma 3.1 are defined, and we clearly know the mapping

$$V \times \cdots \times V \to A(V), \qquad (v_1, \dots, v_p) \mapsto v_1 \wedge \cdots \wedge v_p,$$

is multilinear. Namely, the following formula holds:

$$\begin{aligned} &v_1 \wedge \cdots \wedge (\alpha v_i + \beta v_i') \wedge \cdots \wedge v_p \\ &\quad = \alpha v_1 \wedge \cdots \wedge v_i \wedge \cdots \wedge v_p + \beta v_1 \wedge \cdots \wedge v_i' \wedge \cdots \wedge v_p \\ &\qquad\qquad (\text{where } v_i, v_i' \in V,\ \alpha, \beta \in k). \end{aligned} \tag{3.1}$$

EXAMPLE 3.3. Let $\{i_1, \dots, i_p\}$ be a permutation of $\{1, \dots, p\}$.

$$v_{i_1} \wedge \cdots \wedge v_{i_p} = \operatorname{sgn}\left(\begin{pmatrix} 1 \cdots p \\ i_1 \cdots i_p \end{pmatrix}\right)(v_1 \wedge \cdots \wedge v_p). \tag{3.2}$$

PROOF. Set $\sigma = \binom{1 \cdots p}{i_1 \cdots i_p}$. Then, $v_{i_1} \wedge \cdots \wedge v_{i_p} = \mathscr{A}(v_{i_1} \otimes \cdots \otimes v_{i_p}) = \mathscr{A}(v_{\sigma(1)} \otimes \cdots \otimes v_{\sigma(p)}) = \mathscr{A}(P_{\sigma^{-1}}(v_1 \otimes \cdots \otimes v_p)) = \operatorname{sgn}(\sigma)\mathscr{A}(v_1 \otimes \cdots \otimes v_p) = \operatorname{sgn}(\sigma)(v_1 \wedge \cdots \wedge v_p)$. □

The next proposition is useful for carrying out computations in $A(V)$.

PROPOSITION 3.2. *Let $t \in A^p(V)$ and $t' \in A^q(V)$.*

$$t' \wedge t = (-1)^{pq} t \wedge t'. \tag{3.3}$$

In particular, $t \wedge t = 0$ for odd p.

PROOF. Since $\wedge$ is bilinear, it suffices to prove the result for $t = \mathscr{A}(v_1 \otimes \cdots \otimes v_p)$, $t' = \mathscr{A}(v_{p+1} \otimes \cdots \otimes v_{p+q})$ $(v_i \in V)$. When $p = q = 1$, we have seen in Example 3.1 that $t' \wedge t = -(t \wedge t')$.

To prove general case, we use this formula repeatedly.

$$\begin{aligned} t' \wedge t &= \mathscr{A}(v_{p+1} \otimes \cdots \otimes v_{p+q}) \wedge \mathscr{A}(v_1 \otimes \cdots \otimes v_p) \\ &= (v_{p+1} \wedge \cdots \wedge v_{p+q}) \wedge (v_1 \wedge \cdots \wedge v_p) \\ &= (-1)^p (v_{p+1} \wedge \cdots \wedge v_{p+q-1}) \wedge (v_1 \wedge \cdots \wedge v_p) \wedge v_{p+q}. \end{aligned}$$

Repeating this process, we have

$$\begin{aligned} &= (-1)^{pq}(v_1 \wedge \cdots \wedge v_p) \wedge (v_{p+1} \wedge \cdots \wedge v_{p+q}) \\ &= (-1)^{pq} t \wedge t'. \quad \square \end{aligned}$$

We proved that $\{\mathscr{A}(t(i_1, \dots, i_p)) | i_1 < \cdots < i_p\}$ is a basis for $A^p(V)$ $(p \le n)$ (Proposition 2.7). The next theorem is a restatement of this.

THEOREM 3.1. *Let V be a k-vector space and let $(e_1, \dots, e_n)$ be a basis for V.*

(1) *The exterior algebra $A(V)$ is a k-vector space of dimension 2^n. 1 and $e_{i_1} \wedge e_{i_2} \wedge \cdots \wedge e_{i_p}$ $(1 \le i_1 < i_2 < \cdots < i_p \le n,\ 1 \le p \le n)$ for a basis for $A(V)$.*

(2) *$A(V)$ is generated as an associative algebra by the unit element* 1 *and by* $\{e_1, \ldots, e_n\}$. *And the following relations hold among these*:

$$e_i \wedge e_i = 0 \qquad (i = 1, \ldots, n),$$
$$e_i \wedge e_j = -e_j \wedge e_i \qquad (i, j = 1, \ldots, n, \ i \neq j).$$

DEFINITION 3.2. The basis of $A(V)$ given in Theorem 3.1(1) is called the *standard basis* of $A(V)$ corresponding to a basis $\mathscr{E} = (e_1, \ldots, e_n)$ of V.

In the following, to simplify the notation we denote standard basis elements as follows.

Every element of the standard basis corresponds to a subset of $I = \{1, \ldots, n\}$. Thus, for a subset $H = \{i_1, \ldots, i_p\}$ of I, we denote

$$e_H = e_{i_1} \wedge e_{i_2} \wedge \cdots \wedge e_{i_p},$$

where we assume that $i_1 < i_2 < \cdots < i_p$. We also denote $e_\varnothing = 1$, where $\varnothing$ is the empty set.

COROLLARY TO THEOREM 3.1. *For subsets H and K of I, define the number $\rho_{H,K}$ as follows*;

$$\rho_{H,K} = \begin{cases} 0 & \text{if } H \cap K \neq \varnothing, \\ (-1)^\nu & \text{if } H \cap K = \varnothing, \end{cases} \tag{3.4}$$

where $\nu = \#\{(\lambda, \mu) \in H \times K \mid \lambda > \mu\}$ (2).

Then, the product of standard basis elements is given by

$$e_H \wedge e_K = \rho_{H,K} e_{H \cup K}. \tag{3.5}$$

PROOF. If $H \cap K \neq \varnothing$, $e_H \wedge e_K = 0$ by Lemma 2.2. In this case, $\rho_{H,K} = 0$ by definition, and (3.5) is proved. Assume that $H \cap K = \varnothing$. Let $H = \{\lambda_1, \ldots, \lambda_h\}$ $(\lambda_1 < \cdots < \lambda_h)$ and $K = \{\mu_1, \ldots, \mu_k\}$ $(\mu_1 < \cdots < \mu_k)$. Then $e_H = e_{\lambda_1} \wedge \cdots \wedge e_{\lambda_h}$ and $e_K = e_{\mu_1} \wedge \cdots \wedge e_{\mu_k}$. Though we have $H \cup K = \{\lambda_1, \ldots, \lambda_h, \mu_1, \ldots, \mu_k\}$, we must rearrange the elements in increasing order to get $e_{H \cup K}$. By using the formula $e_i \wedge e_j = -e_j \wedge e_i$ when $i > j$, we exchange the order of terms in

$$(e_{\lambda_1} \wedge \cdots \wedge e_{\lambda_h}) \wedge (e_{\mu_1} \wedge \cdots \wedge e_{\mu_k})$$

ν times. We then have $e_{H \cup K}$. □

For the direct sum $\bigoplus_{p=0}^n A^p(V^*)$ of the spaces $A^p(V^*)$ of alternating covariant tensors, we analogously define the exterior product. Sometimes elements of $A^p(V)$ are called *p-vectors* and elements of $A^p(V^*)$ are called *p-forms*.

By the universal property of tensor algebras (Theorem 2.2) and the properties of the ideal $I(V)$, we obtain the following theorem concerning the universality of exterior algebras.

(2) $\#(X)$ denotes the number of elements of a set X.

THEOREM 3.2. *Let A be an associative algebra with unit element e and let F be a linear mapping of V into A such that $(F(v))^2 = 0$ for all $v \in V$. Then F extends uniquely to an associative algebra homomorphism $\widetilde{F}^{(A)}$ from the exterior algebra $A(V)$ into A. That is, there is a unique associative algebra homomorphism $\widetilde{F}^{(A)}: A(V) \to A$ such that $\widetilde{F}^{(A)}(1) = e$ and $\widetilde{F}^{(A)} \circ \iota_A = F$, where ι_A is the natural inclusion mapping of $V = A^1(V)$ into $A(V)$.*

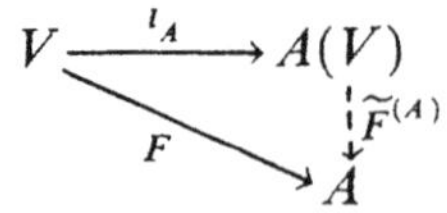

PROOF. By Theorem 2.2, there is an associative algebra homomorphism $\widetilde{F}$ from $T(V)$ into A corresponding to F. If we show that $\widetilde{F}$ equals 0 on the ideal $I(V)$, i..e, $\operatorname{Ker}\widetilde{F} \supseteqq I(V)$ where $\operatorname{Ker}\widetilde{F}$ denotes the kernel of $\widetilde{F}$, then we know that $\widetilde{F}$ naturally induces an (associative algebra) homomorphism $\bar{\widetilde{F}}: T(V)/I(V) \to A$.

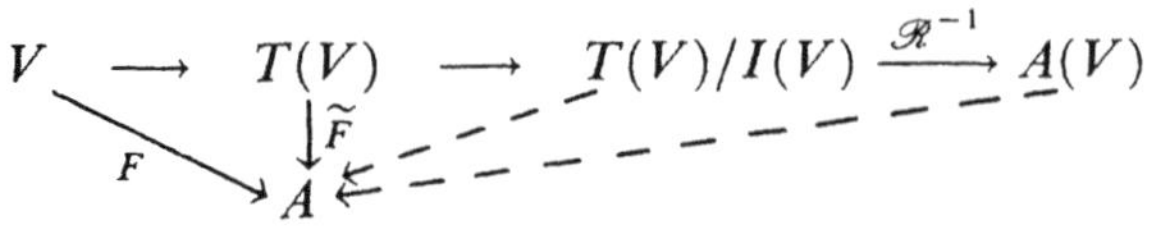

Then the composite mapping $\widetilde{F}^{(A)} = \bar{\widetilde{F}} \circ \mathscr{R}: A(V) \to A$ is a homomorphism. For $t \in A(V)$

$$\widetilde{F}^{(A)}(t) = \bar{\widetilde{F}}(\mathscr{R}(t)) = \bar{\widetilde{F}}(\pi(t)) = \widetilde{F}(t),$$

which implies that $\widetilde{F}^{(A)}$ is the restriction of $\widetilde{F}$ to $A(V)$. In particular, $\widetilde{F}^{(A)}(v) = F(v)$ for $v \in V$.

Now, for all $v \in V$, $\widetilde{F}(v \otimes v) = \widetilde{F}(v) \cdot \widetilde{F}(v) = (F(v))^2 = 0$, which implies that the set $\{v \otimes v | v \in V\}$ of generators of $I(V)$ is included in $\operatorname{Ker}\widetilde{F}$. Since $\operatorname{Ker}\widetilde{F}$ is an ideal, we can conclude that $\operatorname{Ker}\widetilde{F} \supseteqq I(V)$.

Uniqueness follows from the fact that $\widetilde{F}^{(A)}$ is determined by F on the set $\{1\} \cup V$ of generators of $A(V)$. □

COROLLARY. *Let V, W be k-vector spaces and F a linear mapping $V \to W$. Then F extends uniquely to an associative algebra homomorphism from $A(V)$ into $A(W)$.*

PROOF. Let j_A be the natural inclusion mapping of W into $A(W)$. Consider $j_A \circ F: V \to A(W)$. For $v \in V$, $((j_A \circ F)(v))^2 = F(v) \wedge F(v)$ (in $A(W)) = 0$. Thus, we can apply the theorem to $j_A \circ F$. □

We also have the following theorem which characterizes $A(V)$ by its universal property.

THEOREM 3.3. (1) *For a k-vector space V, there exists an associative algebra E_0 with unit element e_0 and a linear mapping ι_0 of V into E_0, which satisfy the following condition* (A).

(A) *Let F be a linear mapping of V into an associative algebra E_1 with unit element e_1 such that $(F(v))^2 = 0$ for all $v \in V$. Then there is a unique homomorphism of associative algebras $\widetilde{F}: E_0 \to E_1$ such that $\widetilde{F} \circ \iota_0 = F$.*

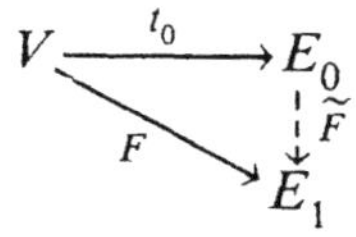

(2) *The pair (E_0, ι_0) is unique in the following sense: If the pairs (E_0, ι_0) and (E_0', ι_0') consisting of an associative algebra and a linear mapping have property (A), then there is a unique isomorphism of associative algebras $\widetilde{F}_0: E_0 \to E_0'$ such that $\widetilde{F}_0 \circ \iota_0 = \iota_0'$.*

PROOF. The proof is similar to that of Theorem 2.3. For part (1), set $E_0 = A(V)$, $\iota_0 = \iota_A$ and apply Theorem 3.2.

COROLLARY. *Let V be a k-vector space of dimension n. Let E be an associative algebra generated by n elements $\{a_1, \dots, a_n\}$ and the unit element e. Moreover, suppose that the following relations hold among the a_i:*

$$
\begin{aligned}
a_i \cdot a_i &= 0 & &(i = 1, \dots, n), \\
a_i \cdot a_j &= -a_j \cdot a_i & &(i, j = 1, \dots, n,\ i \neq j).
\end{aligned}
$$

Take a basis $(e_1, \dots, e_n)$ of V. Then there exists a unique associative algebra homomorphism $\widetilde{F}: A(V) \to E$ such that $\widetilde{F}(e_i) = a_i$ $(i = 1, \dots, n)$. The mapping $\widetilde{F}$ is surjective.

PROOF. Since $(e_1, \dots, e_n)$ is a basis of V, we can define a linear mapping $F: V \to E$ by $F(e_i) = a_i$ $(i = 1, \dots, n)$. Then, for $V \ni v = \sum_{i=1}^n \alpha_i e_i$ $(\alpha_i \in k)$, $F(v) = \sum_{i=1}^n \alpha_i a_i$ and

$$
\begin{aligned}
(F(v))^2 &= \left(\sum_i \alpha_i a_i\right) \cdot \left(\sum_i \alpha_i a_i\right) \\
&= \sum_{i=1}^n \alpha_i^2 a_i \cdot a_i + \sum_{1 \le i < j \le n} \alpha_i \alpha_j (a_i \cdot a_j + a_j \cdot a_i) \\
&= 0.
\end{aligned}
$$

Therefore F extends to a map $\widetilde{F}: A(V) \to E$. □

§2. Applications to determinants

Let $X = (\xi_{ij})$ be an $n \times n$ matrix. The determinant $\det(X)$ of X is defined by

$$\det(X) = \sum_{\sigma \in \mathfrak{S}_n} \operatorname{sgn}(\sigma) \xi_{\sigma(1)1} \xi_{\sigma(2)2} \cdots \xi_{\sigma(n)n}. \tag{3.6}$$

In this section we discuss the connection between exterior algebras and determinants and explain how several properties of determinants are easily deduced using exterior algebras.

First, we introduce some notation. Consider an $n \times m$ matrix $X = (\xi_{ij})$. Let H and K be subsets with r elements of $I = \{1, 2, \dots, n\}$ and $J = \{1, 2, \dots, m\}$ respectively. Arranging the elements of H and K in increasing order, we write $H = \{h_1, \dots, h_r\}$ $(h_1 < \cdots < h_r)$ and $K = \{k_1, \dots, k_r\}$ $(k_1 < \cdots < k_r)$. For $X = (\xi_{ij})$, define

$$X_{H,K} = \begin{bmatrix} \xi_{h_1k_1} & \cdots & \xi_{h_1k_r} \\ \vdots & & \vdots \\ \xi_{h_rk_1} & \cdots & \xi_{h_rk_r} \end{bmatrix},$$

$X_{H,K}$ is an $r \times r$ matrix and $\det(X_{H,K})$ $(H \subseteqq I,\ K \subseteqq J)$ are the r-degree minors of X.

Let $\mathscr{E} = (e_1, \dots, e_n)$ be a basis of a vector space V of dimension n. As before, for a subset $H = \{h_1, \dots, h_r\}$ $(h_1 < \cdots < h_r)$ of $I = (1, \dots, n\}$, define $e_H = e_{h_1} \wedge \cdots \wedge e_{h_r}$.

Proposition 3.3. *Let V be a k-vector space of dimension n and $(e_1, \dots, e_n)$ a basis of V. For $x_1, \dots, x_p \in V$, set*

$$x_i = \sum_{j=1}^{n} \xi_{ji} e_j \qquad (i = 1, \dots, p).$$

Then

$$x_1 \wedge \cdots \wedge x_p = \sum_{\substack{H \subseteqq I, \\ \#(H) = p}} \det(X_{H,L}) e_H, \tag{3.7}$$

where $X = (\xi_{ij})$ and $L = \{1, \dots, p\}$. $\#(H)$ denotes the number of the elements of a set H.

In particular, if $p = n$,

$$x_1 \wedge \cdots \wedge x_p = \det(X)(e_1 \wedge \cdots \wedge e_n). \tag{3.8}$$

Proof. Expand the right side of

$$x_1 \wedge \cdots \wedge x_p = \left(\sum_{i_1} \xi_{i_1 1} e_{i_1}\right) \wedge \left(\sum_{i_2} \xi_{i_2 2} e_{i_2}\right) \wedge \cdots \wedge \left(\sum_{i_p} \xi_{i_p p} e_{i_p}\right)$$

and rewrite it. (Use (3.1) and (3.2).) □

The above formula (3.8) has been obtained using the definition of determinant which was previously given (by (3.6)). Conversely we can define the determinant of $n \times n$ matrix $X = (\xi_{ij})$ by (3.8) as a function of column vectors $\mathbf{x}_1, \ldots, \mathbf{x}_n$ of X. Namely, set

$$\mathbf{x}_i = \begin{bmatrix} \xi_{1i} \\ \xi_{2i} \\ \vdots \\ \xi_{ni} \end{bmatrix} \qquad (i = 1, \ldots, n),$$

$$\mathbf{e}_j = \begin{bmatrix} 0 \\ \vdots \\ 0 \\ 1 \\ 0 \\ \vdots \\ 0 \end{bmatrix} < j \qquad (j = 1, \ldots, n),$$

to form a basis of the k-vector space k^n consisting of column vectors of dimension n. Consider the space $A^n(k^n)$ of alternating tensors of degree n. Since this space is of dimension 1 and $(\mathbf{e}_1 \wedge \cdots \wedge \mathbf{e}_n)$ is a basis, there is a unique element λ of k such that

$$\mathbf{x}_1 \wedge \cdots \wedge \mathbf{x}_n = \lambda(\mathbf{e}_1 \wedge \cdots \wedge \mathbf{e}_n).$$

Using λ, we can define

$$\lambda = \det(\mathbf{x}_1, \ldots, \mathbf{x}_n) = \det(X).$$

From this point of view, (3.8) shows that the determinant thus defined can be computed by the usual formula (3.6). And, such properties of the determinant as multilinearlity and alternativity with respect to columns are immediately obtained by this definition and the corresponding properties of the exterior product $\wedge$.

REMARK. A characterization of the determinant which asserts that n-multilinear and alternating functions of the column vectors $\mathbf{x}_1, \ldots \mathbf{x}_n$ are constant multiples of the determinant, is obtained from the fact that $\dim A^n((k^n)^*) = 1$ (see §2.3(e)).

By the corollary to Theorem 3.2, we have an associative algebra endomorphism $A(F)$ of $A(V)$ corresponding to a linear transformation F of V. We know that $A(F)$ has the following properties:

(1°) Since $A(F)$ is an associative algebra homomorphism, we have, for $v_1, \ldots, v_p \in V$,

$$A(F)(v_1 \wedge \cdots \wedge v_p) = A(F)(v_1) \wedge \cdots \wedge A(F)(v_p) = F(v_1) \wedge \cdots \wedge F(v_p). \tag{3.9}$$

In particular, $A(F)$ maps the space $A^p(V)$ of alternating tensors of degree p into itself. Therefore the restriction of $A(F)$ to $A^p(V)$ is a mapping of $A^p(V)$ into $A^p(V)$, which we denote by $A^p(F)$.

(2°) Let F_1 and F_2 be linear transformations of V. Then, for the composite mapping $F_1 \circ F_2$, we have

$$A(F_1 \circ F_2) = A(F_1) \circ A(F_2).$$

This relation is obtained from the uniqueness of $A(F)$.

Now, we apply these assertions to determinants. As before, denote by $\mathbf{x}_1, \dots, \mathbf{x}_n$ the column vectors of a matrix X. Define a linear transformation F of k^n by $F(\mathbf{e}_i) = \mathbf{x}_i$ $(i = 1, \dots, n)$ Then, by (1°) and Proposition 3.3, we have, for $K \subseteqq I = \{1, \dots, n\}$

$$A(F)(\mathbf{e}_K) = \sum_{\substack{H \subseteqq I \\ \#(H)=\#(K)}} \det(X_{H,K})\mathbf{e}_H. \tag{3.10}$$

In particular, for $K = I$, $A^n(F)$ is a linear transformation of 1-dimensional space $A^n(k^n)$. And when we identify it with a scalar, we get $\det(X)$. In view of this remark, the formula for the determinant of the product matrix XY of X and Y,

$$\det(XY) = \det(X)\det(Y)$$

is obtained immediately from (2°) above. In fact, let $\mathbf{y}_1, \dots, \mathbf{y}_n$ be the column vectors of an $n \times n$ matrix $Y = (\eta_{ij})$, and define a linear transformation G of k^n by $G(\mathbf{e}_i) = \mathbf{y}_i$ $(i = 1, \dots, n)$. By (2°), $A^n(F \circ G) = A^n(F) \circ A^n(G)$. On the other hand, since

$$\begin{aligned}(F \circ G)(\mathbf{e}_i) = F(G(\mathbf{e}_i)) = F(\mathbf{y}_i) = F\left(\sum_k \eta_{ki}\mathbf{e}_k\right) \\ = \sum_k \eta_{ki}\mathbf{x}_k = X\mathbf{y}_i,\end{aligned}$$

$X\mathbf{y}_i$ $(i = 1, \dots, n)$ are the column vectors of XY. Thus $A^n(F \circ G)$ can be regarded as $\det(XY)$.

Next, we shall deduce the Laplace expansion of the determinant from Proposition 3.3. We use the same notation as before. Define a linear transformation F on k^n by $F(\mathbf{e}_i) = \mathbf{x}_i$, where $\mathbf{x}_i$ $(i = 1, \dots, n)$ are the column vectors of an $n \times n$ matrix $X = (\xi_{ij})$. Let us denote by $\mathbf{e}_H = \mathbf{e}_{h_1} \wedge \cdots \wedge \mathbf{e}_{h_r}$ a standard basis element of $A(k^n)$ corresponding to the basis $(\mathbf{e}_1, \dots, \mathbf{e}_n)$ of k^n, where $H = \{h_1, \dots, h_r\}$ $(h_1 < \cdots < h_r)$ is a subset of $I = \{1, \dots, n\}$. Also we write

$$H' = I - H.$$

Lemma 3.2. *Let H and K be subsets of I, such that $\#(H) = r$, $\#(K) = n - r$. Then*

$$A^r(F)(\mathbf{e}_H) \wedge A^{n-r}(F)(\mathbf{e}_K) = \sum_{\substack{R \subseteqq I \\ \#(R)=r}} \rho_{R,R'} \det(X_{R,H}) \det(X_{R',K})\mathbf{e}_I, \tag{3.11}$$

where $\rho_{R,R'}$ is defined by (3.4).

PROOF. By (3.9) and (3.7) we have

$$A^r(F)(\mathbf{e}_H) = F(\mathbf{e}_{h_1}) \wedge \cdots \wedge F(\mathbf{e}_{h_r}) = \mathbf{x}_{h_1} \wedge \cdots \wedge \mathbf{x}_{h_r} = \sum_{\substack{R \subseteqq I \\ \#(R)=r}} \det(X_{R,H})\mathbf{e}_R.$$

Similarly

$$A^{n-r}(F)(\mathbf{e}_K) = \sum_{\substack{S \subseteqq I \\ \#(S)=n-r}} \det(X_{S,K})\mathbf{e}_S.$$

Next we apply the exterior product to these two equalities. Unless $S = R'$ $(= I - R)$, $R \cap S \neq \varnothing$ and $\mathbf{e}_R \wedge \mathbf{e}_S = 0$. Otherwise, $\mathbf{e}_R \wedge \mathbf{e}_{R'} = \rho_{R,R'}\mathbf{e}_I$. □

PROPOSITION 3.4. *Let X be an $n \times n$ matrix and $I = \{1, \dots, n\}$.*
(1) *Let $H \subseteqq I$ and denote $\#(H) = r$. Then,*

$$\det(X) = \rho_{H,H'} \sum_{\substack{R \subseteqq I \\ \#(R)=r}} \rho_{R,R'} \det(X_{R,H}) \det(X_{R',H'}). \tag{3.12}$$

(2) *Let H, $K \subseteqq I$. If $\#(H) = r$, $\#(K) = n - r$, and $H \cap K \neq \varnothing$. Then*

$$\sum_{\substack{R \subseteqq I \\ \#(R)=r}} \rho_{R,R'} \det(X_{R,H}) \det(X_{R',H}) = 0. \tag{3.13}$$

PROOF. In (3.11) substitute $K = H'$. Then the left side becomes

$$A^n(F)(\mathbf{e}_H \wedge \mathbf{e}_{H'}) = A^n(F)(\rho_{H,H'}\mathbf{e}_I) = \rho_{H,H'} \det(X)\mathbf{e}_I.$$

Comparing the coefficients, (1) is obtained. Moreover, if $H \cap K \neq \varnothing$, then the left side of (3.11) is equal to 0, which gives (2). □

COROLLARY (Laplace expansion). *Let $K = \{k_1, \dots, k_r\}$ be a subset of I. Set $\overline{K} = \sum_{i=1}^r k_i$. For subsets R, K of I such that $\#(R) = \#(K)$, set*

$$\widetilde{X}_{R,K} = (-1)^{\overline{R}+\overline{K}} \det(X_{R',K'}).$$

Then, for subsets H, K of I such that $\#(H) = \#(K) = r$,

$$\sum_{\substack{R \subseteqq I \\ \#(R)=r}} \det(X_{R,H})\widetilde{X}_{R,K} = \begin{cases} \det(X) & (\text{if } H = K), \\ 0 & (\text{if } H \neq K). \end{cases} \tag{3.14}$$

PROOF. Assume that $H = K$. From a comparison with Proposition 3.4 (1) it suffices to prove that

$$\rho_{H,H'}\rho_{R,R'} = (-1)^{\overline{R}+\overline{H}}.$$

In general we have

$$\rho_{R,R'} = (-1)^{\overline{R}-r(r+1)/2}.$$

In fact, if $R = \{t_1, \dots, t_r\}$ $(t_1 < \cdots < t_r)$, then the number ν defined by $\nu = \#\{(\lambda, \mu) \in R \times R' | \lambda > \mu\}$ is equal to $\nu = \overline{R} - r(r+1)/2$, since, for $\lambda = t_i$, there exist $(t_i - i)$ elements of R' which are smaller than λ. Therefore

$$\rho_{H,H'}\rho_{R,R'} = (-1)^{\overline{H}+\overline{R}-r(r+1)} = (-1)^{\overline{H}+\overline{R}}.$$

When $H \neq K$, the result follows similarly from Proposition 3.4(2). □

Consider the case $r = 1$ in the corollary. Let $K = \{k\}$, $R = \{j\}$, $H = \{h\}$. Then,

$$\widetilde{X}_{R,K} = (-1)^{k+j} \det(X_{I-\{j\},I-\{k\}}).$$

And (3.14) gives

$$\sum_{j=1}^{n} \xi_{jh}(-1)^{k+j} \det(X_{I-\{j\},I-\{k\}}) = \begin{cases} \det(X) & (h = k), \\ 0 & (h \neq k). \end{cases}$$

These are well-known expansions along columns.

§3. Inner (interior) products of exterior algebras

The space $T_p(V) = T^p(V^*)$ of covariant tensors of degree p can be regarded naturally as the dual space of $T^p(V)$ (see Example 2.3, §2.1). In this case, the inner product (a nondegenerate bilinear form by which these two spaces can be regarded as dual space of each other) is given by, for $T_p(V) \ni \varphi_1 \otimes \cdots \otimes \varphi_p$ and $T^p(V) \ni v_1 \otimes \cdots \otimes v_p$,

$$\langle v_1 \otimes \cdots \otimes v_p, \varphi_1 \otimes \cdots \otimes \varphi_p\rangle = \varphi_1(v_1) \cdots \varphi_p(v_p) \qquad (v_i \in V,\ \varphi_j \in V^*).$$

Using this, we show that the exterior algebras $A(V)$ and $A(V^*)$ can also be regarded as dual spaces of each other. First we define a bilinear form $\langle\ |\ \rangle_p$ on $A^p(V) \times A^p(V^*)$ by

$$\langle z|\xi\rangle_p = p!\langle z, \xi\rangle \qquad (z \in A^p(V),\ \xi \in A^p(V^*)).$$

PROPOSITION 3.5. (1) *For* $z = v_1 \wedge \cdots \wedge v_p$ $(v_i \in V)$, $\xi = \varphi_1 \wedge \cdots \wedge \varphi_p$ $(\varphi_i \in V^*)$, *we have*

$$\langle z|\xi\rangle_p = \det(\varphi_i(v_j)).$$

(2) $A^p(V)$ *and* $A^p(V^*)$ *are dual spaces of each other with respect to the inner product* $\langle\ |\ \rangle_p$. $\{e_H | H \subseteqq I,\ \#(H) = p\}$ *and* $\{f^K | K \subseteqq I,\ \#(K) = p\}$ *are dual bases of each other for* $A^p(V)$ *and* $A^p(V^*)$ *respectively, where* e_H *and* f^K *are standard basis elements of* $A(V)$ *and* $A(V^*)$ *corresponding to a basis* $\mathscr{E} = (e_1, \dots, e_n)$ *of* V *and its dual basis* $\mathscr{F} = (f^1, \dots, f^n)$ *respectively.*

(3) *By using the sum of* $\langle\ |\ \rangle_p$ *as the inner product,* $A(V)$ *and* $A(V^*)$ *are dual spaces of each other. That is, we define a bilinear form* $\langle\ |\ \rangle$ *on*

$A(V) \times A(V^*)$ *as follows, for* $z = \sum_{p=0}^n z_p$ $(z_p \in A^p(V))$ *and* $\xi = \sum_{p=0}^n \xi_p$ $(\xi_p \in A(V^*))$,

$$\langle z|\xi\rangle = \sum_{p=0}^{n} \langle z_p|\xi_p\rangle_p.$$

PROOF. (1) Let $z = v_1 \wedge \cdots \wedge v_p = \mathscr{A}_p(v_1 \otimes \cdots \otimes v_p)$, $\xi = \varphi_1 \wedge \cdots \wedge \varphi_p = \mathscr{A}_p^*(\varphi_1 \otimes \cdots \otimes \varphi_p)$. By $P_\sigma^* = {}^tP_\sigma^{-1}$ (§2.3(e)), $\mathscr{A}_p^* = {}^t\mathscr{A}_p$ is obtained. Thus

$$\begin{aligned}
\langle z|\xi\rangle_p &= p!\langle \mathscr{A}_p(v_1 \otimes \cdots \otimes v_p), \mathscr{A}_p^*(\varphi_1 \otimes \cdots \otimes \varphi_p)\rangle \\
&= p!\langle \mathscr{A}_p(v_1 \otimes \cdots \otimes v_p), \varphi_1 \otimes \cdots \otimes \varphi_p\rangle \\
&= \left\langle \sum_{\sigma \in \mathfrak{S}_p} \operatorname{sgn}(\sigma) v_{\sigma^{-1}(1)} \otimes \cdots \otimes v_{\sigma^{-1}(p)}, \varphi_1 \otimes \cdots \otimes \varphi_p \right\rangle \\
&= \sum_{\sigma \in \mathfrak{S}_p} \operatorname{sgn}(\sigma) \varphi_1(v_{\sigma^{-1}(1)}) \cdots \varphi_p(v_{\sigma^{-1}(p)}) \\
&= \det(\varphi_i(v_j)).
\end{aligned}$$

(2) follows immediately from the formula for $\langle\ |\ \rangle_p$ given in (1). (3) is obtained by (2). □

The transposes of linear transformations of $A(V)$ and $A(V^*)$ are defined with respect to this inner product.

NOTE. Let U_1 and U_1' be k-vector spaces and T a linear mapping: $U_1 \to U_1'$. Then a linear mapping $T_1\colon U_1'^* \to U_1^*$ is defined by

$$T_1(\varphi)(v) = \varphi(T(v)),$$

for $\varphi \in U_1'^*$ And $v \in U_1$. T_1 is called the transpose of T and is denoted by ${}^tT$. Similarly, let U_1 and U_2 be k-vector space, and suppose that U_1 and U_2 are dual spaces of each other by an inner product $\langle\ |\ \rangle\colon U_1 \times U_2 \to k$. Let S be a linear transformation of U_1. Then there exists a unique linear transformation S' of U_2 satisfying, for all $u_1 \in U_1$, $u_2 \in U_2$

$$\langle u_1|S'(u_2)\rangle = (S(u_1)|u_2\rangle.$$

S' is also called the transpose of S and is denoted by ${}^tS$.

EXERCISE. Let V, W be k-vector spaces, F a linear mapping of V into W. By Corollary to Theorem 3.2, F induces a linear mapping $A(F)\colon A(V) \to A(W)$. By the above proposition, we can identify $A(V)^* = A(V^*)$ and $A(W)^* = A(W^*)$. Then prove that ${}^t(A(F)) = A({}^tF)$.

Moreover, using this, prove that, for an $n \times n$ matrix X, $\det({}^tX) = \det X$.

DEFINITION 3.3. Let ξ be an element of $A(V^*)$. Define a linear transformation $\delta(\xi)$ of $A(V^*)$ by

$$\delta(\xi)\zeta = \xi \wedge \xi \qquad (\zeta \in A(V^*)).$$

Denote the transpose of $\delta(\xi)$ by $\partial(\xi)$, which is called the *interior product* by ξ.[3] That is, $\partial(\xi)$ is a linear transformation of $A(V)$ satisfying, for all $z \in A(V)$, $\zeta \in A(V^*)$,

$$\langle \partial(\xi) z | \zeta \rangle = \langle z | \delta(\xi) \zeta \rangle = \langle z | \xi \wedge \zeta \rangle. \tag{3.15}$$

Then we have the following lemma.

LEMMA 3.3. (1) *For* α, $\beta \in k$, ξ, $\eta \in A(V^*)$,

$$\delta(\alpha\xi + \beta\eta) = \alpha\delta(\xi) + \beta\delta(\eta), \qquad \delta(\xi \wedge \eta) = \delta(\xi) \circ \delta(\eta),$$
$$\partial(\alpha\xi + \beta\eta) = \alpha\partial(\xi) + \beta\partial(\eta), \qquad \partial(\xi \wedge \eta) = \partial(\eta) \circ \partial(\xi).$$

(2) *Let* $\xi \in A^p(V)$ *and* $z \in A^q(V)$. *If* $q \geq p$, $\partial(\xi)z \in A^{q-p}(V)$. *If* $q < p$, *then* $\partial(\xi)z = 0$.

(3) *Let* (e_i) *and* (f^j) *be dual basis of* V *and* V^*. *Then, for a standard basis element* e_H, *we have*

$$\partial(f^k)e_H = \begin{cases} \rho_{\{k\},H-\{k\}} e_{H-\{k\}} & (\textit{when } k \in H), \\ 0 & (\textit{otherwise}). \end{cases} \tag{3.16}$$

In general,

$$\partial(f^K)e_H = \begin{cases} \rho_{K,H-K} e_{H-K} & (\text{when } K \subseteqq H), \\ 0 & (\text{otherwise}), \end{cases} \tag{3.17}$$

where $\rho_{H,K}$ *is defined by* (3.4) (§1).

PROOF. (1) and (2) follow easily from the definition.

Let us prove (3). Let H, $K_0 \subseteqq I = \{1, \dots, n\}$, where $\#(H) = q$ and $\#(K_0) = q - 1$.

$$\begin{aligned} \langle \partial(f^k)e_H | f^{K_0} \rangle &= \langle e_H | f^k \wedge f^{K_0} \rangle \\ &= \langle e_H | \rho_{\{k\},K_0} f^{\{k\}\cup K_0} \rangle \\ &= \begin{cases} \rho_{\{k\},K_0} & (\text{when } H = \{k\} \cup K_0, \text{ that is,} \\ & \quad k \in H \text{ and } K_0 = H - \{k\}), \\ 0 & (\text{otherwise}). \end{cases} \end{aligned}$$

Therefore, if $k \in H$, $\partial(f^k)e_H = \rho_{\{k\},H-\{k\}} e_{H-\{k\}}$ and, if $k \notin H$, $\partial(f^k)e_H = 0$, which implies (3.16). In general, let $K = \{k_1, \dots, k_p\}$ $(k_1 < \cdots < k_p)$. Then

$$\partial(f^K)e_H = \partial(f^{k_1} \wedge \cdots \wedge f^{k_p})(e_H) = \partial(f^{k_p}) \circ \cdots \circ \partial(f^{k_1})(e_H).$$

---

[3] Some authors call this the interior multiplication and some denote $\partial(\xi)z$ by $z \llcorner \xi$, since $A(V)$ is a right $A(V^*)$-module by the map $A(V) \times A(V^*) \ni (z, \xi) \mapsto \partial(\xi)z \in A(L)$, which we can see by Lemma 3.3.

If $K \nsubseteq H$, $\partial(f^K)(e_H) = 0$ by (3.16). Assume $K \subseteq H$, and let us prove (3.17) by induction on p. The case $p = 1$ is followed from (3.16). Assume $p > 1$ and set $K_0 = \{k_1, \ldots, k_{p-1}\}$, then

$$\begin{aligned}\partial(f^K)e_H &= \partial(f^{k_p}) \circ \partial(f^{K_0})(e_H) \\ &= \partial(\dot{f}^{k_p})\rho_{K_0, H-K_0} e_{H-K_0} \quad \text{(by induction)} \\ &= \rho_{K_0, H-K_0}\rho_{\{k_p\}, H-K} e_{H-K}.\end{aligned}$$

Since k_p is the largest element of K, we have $\rho_{K_0, H-K_0}\rho_{\{k_p\}, H-K} = \rho_{K, H-K}$, which proves (3.17). □

COROLLARY. *Let $v_1, \ldots, v_q \in V$ and $\varphi_1, \ldots, \varphi_p \in V^*$, where $q \geq p$. Moreover, assume that $\{v_1, \ldots, v_q\}$ is linearly independent. Then*

$$\partial(\varphi_1 \wedge \cdots \wedge \varphi_p)(v_1 \wedge \cdots \wedge v_q) = \sum_{\substack{K \subseteq H \\ \#(K)=p}} \det(X_K)\rho_{K, H-K} v_{H-K}, \tag{3.18}$$

the notation being as follows: Let $H = \{1, \ldots, q\}$, $X = (\varphi_j(v_i))$. For $K = \{k_1, \ldots, k_p\}$ $(k_1 < \cdots < k_p) \subseteq H$, $\det(X_K)$ is the determinant of the $p \times p$ matrix $(\varphi_j(v_i))$, $i \in K$ and $v_M = v_{m_1} \wedge \cdots \wedge v_{m_{q-p}}$ for $M = \{m_1, \ldots, m_{q-p}\}$ $(m_1 < \cdots < m_{q-p})$ $(v_\varnothing = 1)$.

PROOF. Choose a basis $(e_1, \ldots, e_n)$ $(e_i = v_i$ for $i = 1, \ldots, q)$ of V containing $\{v_1, \ldots, v_q\}$. Let $(f^1, \ldots, f^n)$ be the dual basis of $(e_1, \ldots, e_n)$. Then φ_i can be written in the form $\varphi_i = \sum_{j=1}^n \xi_{ji} f^j$, where $\xi_{ji} = \varphi_i(e_j)$ By (3.7) of Proposition 3.3,

$$\varphi_1 \wedge \cdots \wedge \varphi_p = \sum_{\substack{K \subseteq I \\ \#(K)=p}} \det(X_{K,L}) f^K,$$

where $I = \{1, \ldots, n\}$ and $L = \{1, \ldots, p\}$. Therefore

$$\begin{aligned}\partial(\varphi_1 \wedge \cdots \wedge \varphi_p)(v_1 \wedge \cdots \wedge v_q) &= \partial(\varphi_1 \wedge \cdots \wedge \varphi_p)e_H \\ &= \sum_{\substack{K \subseteq I \\ \#(K)=p}} \det(X_{K,L})\partial(f^K)e_H \\ &= \sum_{K \subseteq H} \det(X_{K,L})\rho_{K, H-K} e_{H-K},\end{aligned}$$

which is (3.18). □

We use these formulas in the next section.

§4. Applications to geometry

Let V be an n-dimensional k-vector space and $A^p(V)$ the space of alternating tensors of degree p $(1 \leq p \leq n)$. In this section, we shall examine the correspondence between elements of $A^p(V)$ and subspaces of V.

For $v_1, \ldots, v_p \in V$, the exterior product $v_1 \wedge \cdots \wedge v_p$ is an element of $A^p(V)$, and every element of $A^p(V)$ can be expressed as a linear combination of several elements of this type. For example, taking the standard basis of $A(V)$ corresponding to a basis $\mathscr{E}$ of V, we have such an expression.

First of all, we have the following easy lemma.

LEMMA 3.4. *For $v_1, \ldots, v_p \in V$, $v_1 \wedge \cdots \wedge v_p \neq 0$ if and only if $v_1, \ldots, v_p$ are linearly independent.*

PROOF. If $v_1, \ldots, v_p$ are linearly independent, then there exists a basis for V containing all v_i $(1 \leq i \leq p)$. Then $v_1 \wedge \cdots \wedge v_p$ is a standard basis element of $A^p(V)$. In particular, $v_1 \wedge \cdots \wedge v_p \neq 0$. Conversely, if $v_1, \ldots, v_p$ are linearly dependent, e.g., if $v_1 = \sum_{i=2}^p \alpha_i v_i$, then we have

$$\begin{aligned} v_1 \wedge \cdots \wedge v_p &= \left(\sum_{i=2}^p \alpha_i v_i\right) \wedge v_2 \wedge \cdots \wedge v_p \\ &= \sum_{i=2}^p \alpha_i (v_i \wedge v_2 \wedge \cdots \wedge v_p) = 0. \quad \square \end{aligned}$$

DEFINITION 3.4. An element $z \neq 0$ of $A^p(V)$ which can be expressed in the form $z = v_1 \wedge \cdots \wedge v_p$ with $v_1, \ldots, v_p \in V$, is called *decomposable* or pure.

EXAMPLE 3.4. Nonzero elements of $A^1(V)$ and $A^n(V)$ are decomposable.

EXAMPLE 3.5. Let $e \neq 0$ be an element of $A^n(V)$. Then the mapping of $A^p(V^*)$ into $A^{n-p}(V)$ defined by $\xi \mapsto \partial(\xi)e$ is a linear isomorphism between these two spaces sending decomposable elements to decomposable elements.

In fact, by Definition 3.3 and Lemma 3.3, it is obviously a linear isomorphism. Let $A^p(V^*) \ni \xi = \varphi_1 \wedge \cdots \wedge \varphi_p$ $(\varphi_i \in V^*)$ be decomposable. Then $\varphi_1, \ldots, \varphi_p$ are linearly independent and we can take a basis (f^j) $(f^j = \varphi_i,\ i = 1, \ldots, p)$ of V^* containing them. Let (e_i) be the dual basis of (f^j). e can be expressed in the form $e = \alpha e_1 \wedge \cdots \wedge e_n$, $\alpha \in k$, $\alpha \neq 0$. Therefore $\partial(\xi)e = \pm\alpha e_{p+1} \wedge \cdots \wedge e_n$, which shows that $\partial(\xi)e$ is decomposable.

Using this remark, we know that the nonzero elements of $A^{n-1}(V)$ are decomposable, since every nonzero element of $A^1(V^*)$ is decomposable.

If $v_1, \ldots, v_p$ are linearly independent elements of V, the exterior product $v_1 \wedge \cdots \wedge v_p$ is decomposable by Lemma 3.4. And, if $v_1', \ldots, v_p' \in V$ are linear combinations of $v_1, \ldots, v_p$, then $v_1' \wedge \cdots \wedge v_p'$ is a scalar multiple of $v_1 \wedge \cdots \wedge v_p$, e.g., if $v_i' = \sum_{j=1}^p \alpha_{ji} v_j$, $v_1' \wedge \cdots \wedge v_p' = \det(\alpha_{ij}) v_1 \wedge \cdots \wedge v_p$. Let W be the subspace of V generated by $v_1, \ldots, v_p$. For any basis $(v_1', \ldots, v_p')$ of W, $v_1' \wedge \cdots \wedge v_p'$ is a nonzero scalar multiple of $v_1 \wedge \cdots \wedge v_p$. In fact, we have the next proposition.

PROPOSITION 3.6. *There exists a one-to-one correspondence between the set of p-dimensional subspaces W of V and that of 1-dimensional subspaces L generated by a decomposable element of $A^p(V)$ as follows.*

For a p-dimensional subspace W,

$$L(W) = \{w_1 \wedge \cdots \wedge w_p | w_i \in W\}$$

is a 1-dimensional subspace generated by a decomposable element. On the other hand, for such an L

$$W(L) = \{v \in V | v \wedge z = 0 \text{ for } z \in L, \ z \neq 0\}$$

is a p-dimensional subspace. Moreover, we have

$$W(L(W)) = W, \qquad L(W(L)) = L.$$

In particular, this correspondence is bijective.

PROOF. Choose a basis $(v_1, \ldots, v_p)$ of W, and set $z = v_1 \wedge \cdots \wedge v_p \neq 0$. Then for all $w_1, \ldots, w_p \in W$, $w_1 \wedge \cdots \wedge w_p$ are scalar multiples of z. Therefore $L(W)$ is a 1-dimensional subspace generated by a decomposable element z.

Conversely, let L be a 1-dimensional subspace generated by a decomposable element z. Since z is decomposable, there exist p linearly independent elements $v_1, \ldots, v_p$ of V such that $z = v_1 \wedge \cdots \wedge v_p$. For $v \in V$, $v \wedge z = 0$ if and only if $v, v_1, \ldots, v_p$ are linearly dependent, namely, v belongs to the subspace generated by $v_1, \ldots, v_p$. $W(L)$ is, therefore, the p-dimensional subspace generated by $v_1, \ldots, v_p$.

It is obvious that $W = W(L(W))$. Let z be a nonzero element of L. If $z = v_1 \wedge \cdots \wedge v_p$, then $W(L)$ is the subspace generated by $v_1, \ldots, v_p$. In particular, $z \in L(W(L))$. Comparing the dimensions we have $L = L(W(L))$. □

COROLLARY. *Let $z \in A^p(V)$ be a decomposable element. Suppose that*

$$z = v_1 \wedge \cdots \wedge v_p = v_1' \wedge \cdots \wedge v_p' \qquad (v_i, v_i' \in V),$$

then $v_1', \ldots, v_p'$ are linear combinations of $v_1, \ldots, v_p$.

DEFINITION 3.5. Concerning the correspondence established by Proposition 3.6, the 1-dimensional subspace $L = L(W)$ is called the *Plücker line* corresponding to the subspace W.

If we fix a basis (e_i) for V, a generator z of L can be expressed as $z = \sum_H \xi^H e_H$ using the standard basis for $A(V)$ corresponding to (e_i). Since z is uniquely determined by L up to a nonzero scalar factor, the coefficients $\{\xi^H\}$ are also determined by W up to a nonzero scalar factor. The coefficients are called *Plücker coordinates* of W.

REMARK. L can be considered as a point of the projective space consisting of all 1-dimensional subspaces of $A^p(V)$. Then the Plücker coordinates of W are the homogeneous coordinates of L.

From the corollary to Proposition 3.6, a decomposable element z of $A^p(V)$ determines a p-dimensional subspace. Scalar multiples of z also determine the same space. By fixing an element z we can give more information about W.

EXAMPLE 3.6. Consider the case $k = \mathbf{R}$. Let V be an n-dimensional real vector space and W a p-dimensional subspace of V. Consider the set $\mathcal{F}(W)$ of all bases of W. For $\mathcal{E}$, $\mathcal{E}' \in \mathcal{F}(W)$, $\mathcal{E}$ and $\mathcal{E}'$ are defined to be equivalent if and only if the determinant of the matrix of the change of bases from $\mathcal{E}$ to $\mathcal{E}'$ is positive. Then $\mathcal{F}(W)$ is divided into two equivalence classes. To specify one of them is to choose an *orientation* of W. For example, if $p = 3$, one orientation corresponds to right-handedness and the other to left-handedness. When we fix a generator z for $L(W)$, we can specify one equivalence class as follows. All $\mathcal{E} = (e_1, \dots, e_p) \in \mathcal{F}(W)$ such that $e_1 \wedge \cdots \wedge e_p$ is a positive multiple of z form an equivalence class. Therefore, to fix an element z determines not only a space W but an orientation of W.

PROPOSITION 3.7. *Let W_1 and W_2 be subspaces of a vector space V of dimension p_1 and p_2 respectively. And let L_1 and L_2 be the corresponding Plücker lines ($L_1 \subseteqq A^{p_1}(V)$ and $L_2 \subseteqq A^{p_2}(V)$).*

(1) *W_1 is a subspace of W_2 if and only if there exists a $z \in A^{p_2-p_1}(V)$ such that $L_2 = L_1 \wedge z$.*

(2) *$W_1 \cap W_2 = \{0\}$ is equivalent to $L_1 \wedge L_2 \neq \{0\}$. When this condition is satisfied, the Plücker line of the subspace $W_1 \oplus W_2$ is given by $L_1 \wedge L_2$.*

PROOF. (1) If $W_1 \subseteqq W_2$, there is a basis $(v_1, \dots, v_{p_2})$ for W_2 obtained by extending a basis $(v_1, \dots, v_{p_1})$ for W_1. Then we can choose $z = v_{p_1+1} \wedge \cdots \wedge v_{p_2}$. Conversely, if $L_2 = L_1 \wedge z$ for a $z \in A^{p_2-p_1}(V)$, then $w_1 \wedge L_2 = 0$ for all $w_1 \in W_1$, which implies $W_1 \subseteqq W_2$.

(2) Let $(v_1, \dots, v_{p_1})$ be a basis of W_1, and $(w_1, \dots, w_{p_2})$ that of W_2. $W_1 \cap W_2 = \{0\}$ if and only if $\{v_1, \dots, v_{p_1}, w_1, \dots, w_{p_2}\}$ is linearly independent. For $z_1 = v_1 \wedge \cdots \wedge v_{p_1}$ and $z_2 = w_1 \wedge \cdots \wedge w_{p_2}$, this condition is equivalent to $z_1 \wedge z_2 \neq 0$. If $z_1 \wedge z_2 \neq 0$, $z_1 \wedge z_2$ obviously generates the Plücker line corresponding to $W_1 \oplus W_2$. □

Next, we give a generalization of the correspondence in Proposition 3.6. We associate to a not necessarily decomposable element $z \in A^p(V)$ a subspace of V, which is called the associated space of z. Using this notion, we give a condition under which an element $z \in A^p(V)$ is decomposable.

DEFINITION 3.6. Let $z \in A^p(V)$ $(p \geq 1,\ z \neq 0)$. A subspace W of V is said to *envelope* z if $z \in A^p(W)$, where $A^p(W)$ is regarded naturally as a subspace of $A^p(V)$. In other words, z can be expressed as a sum of exterior products $w_1^{(i)} \wedge \cdots \wedge w_p^{(i)}$ of elements $w_1^{(i)}, \dots, w_p^{(i)}$ of W.

Then we have

PROPOSITION 3.8. *A subspace W envelopes $z \in A^p(V)$ if and only if $\partial(\xi)z \in W$ for all $\xi \in A^{p-1}(V^*)$.*

PROOF. Extend a basis $(e_1, \dots, e_k)$ of W to obtain a basis $(e_1, \dots, e_n)$ for V. Let $(f^1, \dots, f^n)$ be the dual basis of it. By means of the standard basis, z can be expressed uniquely as $z = \sum_{H \subseteqq I} \xi^H e_H$, where $I = \{1, \dots, n\}$ and $\xi^H \in k$. Then, $z \in A^p(W)$ if and only if $\xi^H = 0$ for all $H \nsubseteqq \{1, \dots, k\}$.

On the other hand, since ∂ is linear, the condition stated in the proposition is satisfied if it is true for all standard basis elements f^K for $A^{p-1}(V^*)$, where $K \subseteqq I$ and $\#(K) = p-1$. Suppose that $\#(K) = p-1$. Then, by (3.17), we have

$$\partial(f^K)z = \sum_H \xi^H(\partial(f^K)e_H) = \sum_{\substack{H \\ K \subseteqq H}} \xi^H \rho_{K,H-K} e_{H-K},$$

where $H-K$ consists of one element. As we can easily see, in order to have $\partial(f^K)z \in W$ for all K, it is necessary and sufficient that $\xi^H = 0$ for all $H \nsubseteqq \{1, \dots, k\}$, which implies the proposition. □

This proposition shows that there exists a minimal subspace $V(z)$ which envelopes z and that $V(z)$ is the subspace generated by $\{\partial(\xi)z | \xi \in A^{p-1}(V^*)\}$.

DEFINITION 3.7. $V(z)$ is called the *associated space* of z and the dimension of $V(z)$ is called the *rank* of z, which is denoted by $\operatorname{rank}(z)$. If $z \in A^p(V)$ and $z \neq 0$, it is obvious that $\operatorname{rank}(z) \geq p$.

PROPOSITION 3.9. *A nonzero element $z \in A^p(V)$ is decomposable if and only if* $\operatorname{rank}(z) = p$. *In this case, The Plücker line corresponding to the p-dimensional space $V(z)$ is generated by z.*

PROOF. If $\operatorname{rank}(z) = p$, i.e., $\dim V(z) = p$, then $A^p(V(z))$ is 1-dimensional and $z \in A^p(V(z))$. Thus, using a basis $(e_1, \dots, e_p)$ of $V(z)$, z can be expressed as $z = \alpha e_1 \wedge \cdots \wedge e_p$ $(\alpha \in k)$. Conversely, if z is decomposable, i.e., $z = v_1 \wedge \cdots \wedge v_p$, $V(z)$ is, by definition, the minimal subspace which envelopes z and $V(z)$ is contained in the subspace of V generated by $\{v_1, \dots, v_p\}$. This implies $\operatorname{rank}(z) = \dim V(z) \leq p$. From this and the remark above we have that $\operatorname{rank}(z) = p$. The second part follows from the proof of the first part. □

We can reformulate the condition for decomposability.

PROPOSITION 3.10. *Let $z \in A^p(V)$ with $z \neq 0$. z is decomposable if and only if, for all $\xi \in A^{p-1}(V^*)$,*

$$\partial(\xi)z \wedge z = 0. \tag{3.19}$$

PROOF. Suppose that z is decomposable. Set $z = v_1 \wedge \cdots \wedge v_p$. Then, by (3.18), $\partial(\xi)z$ is a linear combination of v_i. Therefore $\partial(\xi)z \wedge z = 0$.

Conversely, suppose that z is not decomposable, namely, $\operatorname{rank}(z) = k > p$. Let $(e_1, \dots, e_k)$ be a basis of the associated space $V(z)$ of z. Since $V(z)$ is generated by $\partial(\xi)z$ by definition, the e_j can be expressed as $e_j = \partial(\xi_j)z$ where $\xi_j \in A^{p-1}(V^*)$ $(j = 1, \dots, k)$. If $\partial(\xi)z \wedge z = 0$ for all $\xi \in A^{p-1}(V^*)$, then $e_j \wedge z = 0$ for all j. Set $z = \sum_{H \subseteqq \{1, \dots, k\},\ \#(H)=p} \xi^H e_H$. Then, $e_j \wedge z = \sum_{j \notin H}(\pm\xi^H)e_{H\cup\{j\}} = 0$ and all $\xi^H = 0$, which contradicts the assumption that $z \neq 0$. □

If we fix a basis of $A^{p-1}(V^*)$ and assume (3.19) then for each basis element we have the same result as in Proposition 3.19. Thus, choose a basis (e_i) for V and its dual basis (f^j). Using the standard basis, (3.19) can be written as follows: For any $K \subseteqq I$ such that $\#(K) = p-1$ and $L \subseteqq I$ such that $\#(L) = p+1$ (where $I = \{1, \dots, n\}$),

$$\langle \partial(f^K)z \wedge z | f^L \rangle = 0. \tag{3.19 (K, L)}$$

LEMMA 3.5. *Let K, L, J, and J_1 be subsets of I such that $\#(K) = p-1$, $\#(L) = p+1$, $\#(J) = \#(J_1) = p$.*

$$\langle \partial(f^K)e_J \wedge e_{J_1} | f^L \rangle = \begin{cases} \rho_{K,\{i\}}\rho_{\{i\},J_1} & \text{if } J = K \cup \{i\},\ L = \{i\} \cup J_1, \\ 0 & \text{otherwise.} \end{cases}$$

PROOF. The lemma is easily obtained by direct computation. □

Fix i and, for K and L such that $i \notin K$, and $i \in L$ (where $\#(K) = p-1$, $\#(L) = p+1$), define

$$\rho_{K,\{i\}}\rho_{\{i\},L-\{i\}} = \varepsilon_{i,L,K}.$$

By definition (3.4) of ρ, $\varepsilon_{i,L,K}$ is equal to $(-1)^\delta$, where

$$\delta = \#\{l \in L | l < i\} + \#\{k \in K | k > i\}.$$

Take $z \in A^p(V)$, with $z \neq 0$. Write $z = \sum_H \xi^H e_H$, and substitute this into (3.19 (K, L)). Form Lemma 3.5, we have

$$\sum_{i \notin K,\ i \in L} \varepsilon_{i,L,K}\xi^{K\cup\{i\}}\xi^{L-\{i\}} = 0. \tag{3.20 (K, L)}$$

These equalities are called *Plücker relations*. These relations can be regarded as a necessary and sufficient condition for a set $\{\xi^H\}$ of elements of k (not all of them are equal to 0) to be Plücker coordinates of a certain p-dimensional subspace.

EXAMPLE 3.7. The case where $n = 4$, $p = 2$. It is easily seen that, for $K \subsetneqq L$, the equality (3.20 (K, L)) is a trivial equality. By the condition that $\#(L) = 3$ and $\#(K) = 1$, it suffices to consider the following 4 cases: $K = \{i\}$, $L = \{1, 2, 3, 4\} - \{i\}$ $(i = 1, 2, 3, 4)$. For all of them, the equality reduces to the following one.

$$\xi^{\{1,2\}}\xi^{\{3,4\}} - \xi^{\{1,3\}}\xi^{\{2,4\}} + \xi^{\{1,4\}}\xi^{\{2,3\}} = 0.$$

EXAMPLE 3.8. Let

$$z = \sum_{i_1<\cdots<i_p} \xi^{i_1,\dots,i_p}(e_{i_1}\wedge\cdots\wedge e_{i_p}).$$

Then the Plücker relations can be written as follows:

$$\sum_{k=1}^{p+1}(-1)^k \xi^{j_1,\dots,j_{p-1}i_k}\xi^{i_1,\dots,i_{k-1}i_{k+1},\dots,i_{p+1}} = 0$$

$$(j_1 < \cdots < j_{p-1},\ i_1 < \cdots < i_{p+1}),$$

where, for all p indices $l_1, \dots, l_p \in \{1, 2, \dots, n\}$, $\xi^{l_1,\dots,l_p}$ is defined by the next formulas:

(1) If $l_i = l_j$ for at least one pair $i \neq j$, $\xi^{l_1,\dots,l_p} = 0$.

(2) If $l_1, \dots, l_p$ are all different and after arranging them in increasing order, $\{l_1, \dots, l_p\} = \{i_1, \dots, i_p\}$ $(i_1 < \cdots < i_p)$,

$$\xi^{l_1,\dots,l_p} = \operatorname{sgn}\left(\begin{pmatrix} i_1,\dots,i_p \\ l_1,\dots,l_p \end{pmatrix}\right)\xi^{i_1,\dots,i_p}.$$

PROOF. Let $L = \{i_1, \dots, i_{p+1}\}$ and $K = \{j_1, \dots, j_{p-1}\}$. By (3.20 (K, L)), we have

$$\sum_{k=1}^{p+1} \varepsilon_{i_k, L, K}\xi^{K\cup\{i_k\}}\xi^{L-\{i_k\}} = 0,$$

where $\xi^{\{l_1,\dots,l_p\}} = \xi^{l_1,\dots,l_p}(l_1 < \cdots < l_p)$, and $\xi^{K\cup\{i_k\}} = 0$ if $i_k \in K$. Into this formula substitute the following:

$$\varepsilon_{i_k,L,K} = (-1)^{k-1} \times (-1)^{\delta_k} \qquad (\delta_k = \#\{j \in K | j > i_k\}),$$

$$\xi^{K\cup\{i_k\}} = (-1)^{\delta_k}\xi^{j_1,\dots,j_{p-1}i_k} \quad (i_k \notin K). \quad \square$$

REMARK. This result implies that, in the projective space consisting of all 1-dimensional subspaces of $A^p(V)$, the subset of 1-dimensional subspaces generated by decomposable elements is defined by these homogeneous quadratic polynomials. As we have explained, this subset corresponds to the set of p-dimensional subspaces of V.

Exercises

1. Let $V = \mathbf{R}^3$ be the dual vector space consisting of all the column vectors of dimension 3. Let $\mathscr{E} = (\mathbf{e}_1, \mathbf{e}_2, \mathbf{e}_3)$ be the basis of V consisting of the unit vectors (see Exercises in Chapter 1).

   (1) v_1, v_2, and v_3 are elements of V given below.

$$v_1 = \begin{bmatrix} 4 \\ 3 \\ 5 \end{bmatrix}, \qquad v_2 = \begin{bmatrix} 3 \\ 2 \\ -2 \end{bmatrix}, \qquad v_3 = \begin{bmatrix} 6 \\ 2 \\ 3 \end{bmatrix}.$$

Express $v_1 \wedge v_2$ and $v_1 \wedge v_2 \wedge v_3 \in A(V)$ using the standard basis corresponding to $\mathscr{E}$.

(2) Let F be the linear transformation of V whose matrix with respect to $\mathscr{E}$ is given by

$$\begin{bmatrix} 1 & 2 & 3 \\ -1 & 3 & -2 \\ 2 & 5 & 3 \end{bmatrix}.$$

Give the matrices of the linear transformations $A^p(F)$ $(p = 2, 3)$ induced by F, with respect to the standard basis corresponding to $\mathscr{E}$.

(3) We can define an alternating bilinear form on $V \times V$ by $(v, w) \mapsto {}^t vAw \in \mathbf{R}$, where A is given by

$$A = \begin{bmatrix} 0 & 2 & -1 \\ -2 & 0 & 3 \\ 1 & -3 & 0 \end{bmatrix}.$$

When we regard this form as an element of $A^2(V^*)$, express it using the standard basis of $A^2(V^*)$ corresponding to the dual basis of $\mathscr{E}$.

Do the same for any alternating matrix

$$\begin{bmatrix} 0 & \alpha & \gamma \\ -\alpha & 0 & \beta \\ -\gamma & -\beta & 0 \end{bmatrix}.$$

In the following exercises, V denotes a k-vector space.

2. Let $z \in A^p(V)$ and $z' \in A^q(V)$. Consider z and z' as elements of $T^p(V)$ and $T^q(V)$ respectively, and suppose that their components with respect to a basis $\mathscr{E}$ of V are given by $\{\xi^{j_1, \dots, j_p}\}$ and $\{\eta^{k_1, \dots, k_q}\}$. Express the components of $z \wedge z' = \mathscr{A}(z \otimes z') \in T^{p+q}(V)$ with respect to $\mathscr{E}$ using $\xi^{j_1, \dots, j_p}$ and $\eta^{k_1, \dots, k_q}$.
3. Assume that $v_1, \dots, v_p \in V$ are linearly independent. If we have
$$v_1 \wedge w_1 + \cdots + v_p \wedge w_p = 0$$
for $w_1, \dots, w_p \in V$, prove that each w_i can be written in the form $w_i = \sum_{j=1}^p \alpha_{ij} v_j$, where $\alpha_{ij} = \alpha_{ji}$.
4. Let F be a linear transformation of V and (α_{ij}) its matrix with respect to a basis $\mathscr{E}$ of V.

   (1) Compute the matrix of the linear transformation $A^p(F)$ of $A^p(V)$ induced by F with respect to the standard basis corresponding to $\mathscr{E}$.

   (2) If the eigenvalues of F are $\alpha_1, \dots, \alpha_n$ $(n = \dim V)$, prove that the eigenvalues of $A^2(F)$ are $\alpha_i \alpha_j$ $(1 \le i < j \le n)$.

   (3) Express the eigenvalues of $A^p(F)$ using $\alpha_1, \dots, \alpha_n$.

   (4) Express the determinant of $A^p(F)$ using that of F.
5. The mapping $Q: V \to k$ satisfying the following conditions (Q1) and (Q2) is called a quadratic form on V.

(Q1) For all $\alpha \in k$, $v \in V$, $Q(\alpha v) = \alpha^2 Q(v)$.

(Q2) $B(v, w) = Q(v+w) - Q(v) - Q(w)$ is a symmetric bilinear form on $V \times V$.

Then the pair $(\mathscr{C}, \iota)$ consisting of an associative algebra $\mathscr{C}$ with unit element e and a linear mapping ι of V into $\mathscr{C}$ satisfying the conditions (C1) and (C2) below, is called a *Clifford algebra* for Q.

(C1) For all $v \in V$, $(\iota(v))^2 = Q(v) \cdot e$.

(C2) Let A be an associative algebra with the unit element e' and F a linear mapping of V into A such that $(F(v))^2 = Q(v) \cdot e'$ for all $v \in V$. Then there is a unique associative algebra homomorphism $\widetilde{F}$ of $\mathscr{C}$ into A such that $\widetilde{F} \circ \iota = F$.

(1) We know that a Clifford algebra for Q is essentially unique. Formulate this assertion in the form of a proposition analogous to Theorem 2.3 and Theorem 3.3, and prove it.

(2) Let J be the ideal of the tensor algebra $T(V)$ generated by $\{v \otimes v - Q(v) \cdot 1 | v \in V\}$ and let ι be the natural mapping $V \to T(V) \to T(V)/J$. Show that $(T(V)/J, \iota)$ is a Clifford algebra for Q.

(3) For $\dim V = n$, prove that the dimension of a Clifford algebra is at most 2^n. (In fact, it is equal to 2^n.)

6. Let X be an $m \times n$ matrix and Y an $n \times m$ matrix where $m \leq n$. Then we know that

$$\det(XY) = \sum_{\substack{K \subseteq I \\ \#(K)=m}} \det(X_{J,K}) \det(Y_{K,J}),$$

where $I = \{1, \dots, n\}$, $J = \{1, \dots, m\}$.

Prove this formula using exterior algebras. Define a linear mapping $F: k^n \to k^m$ by $\mathbf{y} \mapsto X\mathbf{y}$ ($\mathbf{y} \in k^n$). To get $\det(XY)$, it is enough to compute $F(\mathbf{y}_1) \wedge \cdots \wedge F(\mathbf{y}_m) = A(F)(\mathbf{y}_1 \wedge \cdots \wedge \mathbf{y}_m)$, where $\mathbf{y}_1, \dots, \mathbf{y}_m$ are the column vectors of Y.

7. Let $z \in A^2(V)$, $z \neq 0$. Show that there exists an even number of elements $v_1, \dots, v_{2r}$ of V, which are linearly independent, such that

$$z = v_1 \wedge v_2 + v_3 \wedge v_4 + \cdots + v_{2r-1} \wedge v_{2r}.$$

Show also that the associated space $V(z)$ is generated by $v_1, \dots, v_{2r}$.

8. Let W be a subspace of V and $z \in A(V^*)$. W is said to annihilate z, when $z \circ i = 0$, where i is the inclusion mapping of W into V. In other words, when z is expressed as a sum of alternating forms of different numbers of variables (see §2.3 (e)), the restriction to W of each term is equal to the zero form. Prove the following assertions.

(1) When $z \in A^q(V^*)$ and $\dim W < q$, then W annihilates z.

(2) When $z \in A^p(V^*)$ and $\dim W = p$, let z_W be a nonzero element of the Plücker line corresponding to W. Then W annihilates z if and only if $\langle z_w | z \rangle = 0$.

(3) If a subspace W of V annihilates every element of a subset Σ of $A(V^*)$, then W annihilates every element of the ideal of $A(V^*)$ generated by Σ.

(4) $A(V^*)$ can be considered as graded associative algebra, like $A(V)$. Let I be a homogeneous ideal of $A(V^*)$. When $\dim W = p$, W annihilates every element of I if and only if $\langle z_W | z \rangle = 0$ for all $z \in I \cap A^p(V^*)$, where z_W is defined in (2).

CHAPTER IV

Algebraic Systems with Bilinear Multiplication. Lie Algebras

In Chapter 2, we mentioned that bilinear multiplication is given by a tensor. In this chapter, we shall present some generalities about algebraic systems with bilinear multiplication. Then as an example, we give the definition of Lie algebra and discuss some of its properties. For this purpose, we shall introduce in §2 the theory of replicas of matrices invented by Chevalley. As this theory is an application of the theory of tensor spaces, it seems interesting from this point of view as well.

§1. Algebraic systems with bilinear multiplication

Let A be a k-vector space and m a bilinear mapping m of $A \times A$ into A. As we explained in §2.1 Example 2.4, we can define a multiplication on A by calling $m(a_1, a_2)$ the product of a_1 and a_2. In this case A (more exactly, the pair (A, m)) is called an *algebra* over k. In the following, we sometimes write $m(a_1, a_2) = a_1 \cdot a_2$ for simplicity.

An associative algebra—in §2.4(b)—and a Lie algebra whose definition we give below, are interesting examples of algebras.

For an algebra in general, using the mapping m, we can define homomorphism, isomorphism, subalgebra, ideal, factor algebra, and other properties in a similar fashion as in the case of an associative algebra.

Let A be an algebra of dimension n and $\mathscr{E} = (e_1, \dots, e_n)$ a basis of A. The product $m(e_i, e_j)$ can be expressed as a linear combination of basis elements,

$$m(e_i, e_j) = \sum_{r=1}^{n} \alpha_{ij}{}^r e_r \qquad (\alpha_{ij}{}^r \in k).$$

The n^3 elements $\alpha_{ij}{}^r$ of k are called the *structure constants* of A with respect to the basis $\mathscr{E}$. We shall show below that they are the components of m with respect to $\mathscr{E}$, when m is regarded as tensor of type $(1, 2)$.

By the isomorphism $\mathscr{L}(A, A; A) \cong T_2^1(A)$ explained in Example 2.4, §2.1, to a standard basis element $t_r{}^{pq}$ $(= f^p \otimes f^q \otimes e_r$, where $(f^1, \dots, f^n)$ is the dual basis of $\mathscr{E})$, there corresponds a bilinear mapping of $A \times A$

into A such that $(e_i, e_j) \mapsto f^p(e_i)f^q(e_j)e_r = \delta_{pi}\delta_{qj}e_r$ (where δ_{ij} denotes Kronecker's symbol). Thus, if we express the mapping m by standard basis elements as

$$m = \sum_{p,q,r} \xi_{pq}{}^r t_r{}^{pq} \qquad (\xi_{pq}{}^r \in k),$$

we have

$$m(e_i, e_j) = \sum_{p,q,r} \xi_{pq}{}^r \delta_{pi}\delta_{qj}e_r = \sum_r \xi_{ij}{}^r e_r,$$

which shows $\alpha_{ij}{}^r = \xi_{ij}{}^r$.

Let A be a k-vector space. In order to make A an algebra it is enough to give a mapping m which defines a multiplication. When we fix a basis for A, to give m is equivalent to giving structure constants $\{\alpha_{ij}{}^r\}$. Therefore, in many cases, conditions on multiplication can be described easily as conditions on structure constants. Let us give an example. Associativity means that, for all $a, a', a'' \in A$, $(a\cdot a')\cdot a'' = a\cdot(a'\cdot a'')$. By the bilinearity of m, it suffices to establish the formula for basis elements. Therefore, by rewriting the formula $(e_i\cdot e_j)\cdot e_k = e_i\cdot(e_j\cdot e_k)$, associativity is equivalent to

$$\sum_l \alpha_{ij}{}^l \alpha_{lk}{}^m = \sum_l \alpha_{il}{}^m \alpha_{jk}{}^l,$$

for all i, j, k, m.

Next, we give the definition of Lie algebra, which is the main theme of this chapter.

DEFINITION 4.1. A set $\mathfrak{g}$ is called a *Lie algebra* over k if it satisfies the following conditions.

(1) $(\mathfrak{g}, m)$ is an algebra. That is, $\mathfrak{g}$ is a k-vector space and m is a bilinear mapping $\mathfrak{g}\times\mathfrak{g}\to\mathfrak{g}$.

(2) m satisfies the following conditions (i) and (ii). (As is customary we denote the multiplication in a Lie algebra by $m(X, Y) = [X, Y]$ for $X, Y \in \mathfrak{g}$.)

(i) For all $X \in \mathfrak{g}$, $[X, X] = 0$.

(ii) For all $X, Y, Z \in \mathfrak{g}$,

$$[X, [Y, Z]] + [Y, [Z, X]] + [Z, [X, Y]] = 0.$$

Condition (ii) is called *the Jacobi identity.*

Condition (i) implies that, for all $X, Y \in \mathfrak{g}$, $[X, Y] = -[Y, X]$. In fact, $0 = [X+Y, X+Y] = [X, X]+[X, Y]+[Y, X]+[Y, Y] = [X, Y]+[Y, X]$.

EXERCISE 1. Describe the conditions for $(\mathfrak{g}, m)$ to be a Lie algebra, using the structure constants.

EXAMPLE 4.1. Let (A, m) be an algebra. Denote by $\mathfrak{g}$ the set $\mathrm{Der}(A)$ of derivations of A. (See §2.4(d) for the definition.) $\mathrm{Der}(A)$ is a vector space since it is a subspace of the vector space of linear transformations of

A. Namely, for $D_1, D_2 \in \mathrm{Der}(A)$, the sum $D_1 + D_2$ and the scalar multiple αD_1 ($\alpha \in k$) are defined as follows. For $a \in A$,

$$(D_1 + D_2)(a) = D_1(a) + D_2(a), \qquad (\alpha D_1)(a) = \alpha(D_1(a)).$$

For $D_1, D_2 \in \mathrm{Der}(A)$, the composite mapping $D_1 \circ D_2$ is not necessarily a derivation, but we know that $D_1 \circ D_2 - D_2 \circ D_1 \in \mathrm{Der}(A)$. In fact, (here we write $m(a, a') = a \cdot a'$)

$$\begin{aligned}
&(D_1 \circ D_2 - D_2 \circ D_1)(a \cdot a') \\
&\quad = D_1(D_2(a) \cdot a' + a \cdot D_2(a')) - D_2(D_1(a) \cdot a' + a \cdot D_1(a')) \\
&\quad = (D_1 \circ D_2)(a) \cdot a' + D_2(a) \cdot D_1(a') + D_1(a) \cdot D_2(a') + a \cdot (D_1 \circ D_2)(a') \\
&\qquad -\{(D_2 \circ D_1)(a) \cdot a' + D_1(a) \cdot D_2(a') + D_2(a) \cdot D_1(a') \\
&\qquad\qquad + a \cdot (D_2 \circ D_1)(a')\} \\
&\quad = (D_1 \circ D_2 - D_2 \circ D_1)(a) \cdot a' + a \cdot (D_1 \circ D_2 - D_2 \circ D_1)(a').
\end{aligned}$$

Therefore, if we define $[D_1, D_2] = D_1 \circ D_2 - D_2 \circ D_1$ for $D_1, D_2 \in \mathrm{Der}(A)$, this is a bilinear mapping of $\mathrm{Der}(A) \times \mathrm{Der}(A)$ into $\mathrm{Der}(A)$, and moreover it satisfies the condition (2) of Definition 4.1. So $\mathfrak{g} = \mathrm{Der}(A)$ forms a Lie algebra by this operation.

EXAMPLE 4.2. Let (A, m) be an associative algebra. We denote the product $m(a, a')$ by $a \cdot a'$. Then, define $[a, a'] = a \cdot a' - a' \cdot a$ for $a, a' \in A$. This is a bilinear mapping of $A \times A$ into A, which satisfies condition (2) of Definition 4.1. So A forms a Lie algebra by this multiplication. When $A = M_n(k)$ and $A = \mathrm{Hom}(V, V)$ (cf. Examples 2.15 and 2.15′, §2.4) the Lie algebras thus defined are denoted by $\mathfrak{gl}_n(k)$ and $\mathfrak{gl}(V)$ respectively.

Let $\mathfrak{g}$ be a Lie algebra. A homomorphism, isomorphism, subalgebra, and ideal of $\mathfrak{g}$ as an algebra are called a Lie algebra homomorphism, Lie algebra isomorphism, Lie subalgebra, and ideal, respectively. For an ideal $\mathfrak{a}$, the factor algebra $\mathfrak{g}/\mathfrak{a}$ is defined and, in fact, it is a Lie algebra. For instance, $\mathfrak{a} \subseteqq \mathfrak{g}$ is an ideal of $\mathfrak{g}$ if $\mathfrak{a}$ is a subspace of $\mathfrak{g}$ and $[X, A] \in \mathfrak{a}$ for all $X \in \mathfrak{g}$ and $A \in \mathfrak{a}$. (In this case, $[A, X] = -[X, A]$ is also contained in $\mathfrak{a}$.)

EXAMPLE 4.3. Let $\mathfrak{g}$ be a Lie algebra.

$$\mathfrak{z} = \{X \in \mathfrak{g} | [X, Y] = 0 \text{ for all } Y \in \mathfrak{g}\}$$

is called the *center* of $\mathfrak{g}$. The center of $\mathfrak{g}$ is an ideal.

EXERCISE 2. Let (A, m) be an algebra. Prove that A is a Lie algebra if and only if, for all $a \in A$ the mapping $A \to A$, $a \mapsto m(a, a')$ is a derivation which sends a to 0.

DEFINITION 4.2. Let $\mathfrak{g}$ be a Lie algebra over k and V a k-vector space. A Lie algebra homomorphism Φ of $\mathfrak{g}$ into the Lie algebra $\mathfrak{gl}(V)$ (i.e., a linear mapping $\Phi: \mathfrak{g} \to \mathfrak{gl}(V)$ such that $\Phi([X, Y]) = [\Phi(X), \Phi(Y)] = \Phi(X)\Phi(Y) - \Phi(Y)\Phi(X)$ for all $X, Y \in \mathfrak{g}$) is called a *representation* of $\mathfrak{g}$ on V. Then V is called a *representation space* of $\mathfrak{g}$.

EXAMPLE 4.4. (Contragredient representation). Let $\mathfrak{g}_0 = \mathfrak{gl}(V)$. Consider the dual space V^*. The mapping $\mathfrak{g} \to \mathfrak{gl}(V^*)$ which assigns $-{}^tX \in \mathfrak{gl}(V^*)$ to $X \in \mathfrak{g}_0$ defines a representation of $\mathfrak{g}_0$. In fact,

$$\begin{aligned} -{}^t([X, Y]) &= -{}^t(XY - YX) = {}^tX{}^tY - {}^tY{}^tX \\ &= (-{}^tX)(-{}^tY) - (-{}^tY)(-{}^tX), \end{aligned}$$

which shows that this mapping is a Lie algebra homomorphism.

Generally, let Φ be a representation of $\mathfrak{g}$ on V. The mapping which assigns $-{}^t\Phi(X) \in \mathfrak{gl}(V^*)$ to $X \in \mathfrak{g}$ defines a representation of $\mathfrak{g}$ on V^*. This is called the *contragradient representation* of $\mathfrak{g}$.

EXAMPLE 4.5. (Tensor product of representations). Let $\Phi_i\colon \mathfrak{g} \to \mathfrak{gl}(V_i)$ $(i = 1, 2)$ be representations of a Lie algebra $\mathfrak{g}$. Define a mapping $\Phi_1 \otimes \Phi_2\colon \mathfrak{g} \to \mathfrak{gl}(V_1 \otimes V_2)$ by

$$(\Phi_1 \otimes \Phi_2)(X) = \Phi_1(X) \otimes I_{V_2} + I_{V_1} \otimes \Phi_2(X)$$

for $X \in \mathfrak{g}$, where I_{V_i} denotes the identity transformation of $V_i = \mathrm{id}_{V_i}$. Then $\Phi_1 \otimes \Phi_2$ defines a representation of $\mathfrak{g}$ on $V_1 \otimes V_2$, which is called the *tensor product representation* of Φ_1 and Φ_2.

EXAMPLE 4.6. (Adjoint representation). Let $\mathfrak{g}$ be a Lie algebra and X an element of $\mathfrak{g}$. The mapping which assigns $[X, Y] \in \mathfrak{g}$ to $Y \in \mathfrak{g}$ is obviously a linear transformation of $\mathfrak{g}$, which is denoted by $\mathrm{ad}(X)$. The mapping $\mathrm{ad}\colon \mathfrak{g} \to \mathfrak{gl}(V)$ which assigns $\mathrm{ad}(X) \in \mathfrak{gl}(\mathfrak{g})$ to $X \in \mathfrak{g}$ is not only linear, but also a Lie algebra homomorphism due to the Jacobi identity:

$$\begin{aligned} \mathrm{ad}([X, Y])(Z) &= [[X, Y], Z] = [X, [Y, Z]] - [Y, [X, Z]] \\ &= \mathrm{ad}(X)\,\mathrm{ad}(Y)(Z) - \mathrm{ad}(Y)\,\mathrm{ad}(X)(Z), \qquad X, Y, Z \in \mathfrak{g}. \end{aligned}$$

Thus, the mapping ad is a representation of a Lie algebra $\mathfrak{g}$ with representation space $\mathfrak{g}$, which is called the *adjoint representation* of $\mathfrak{g}$.

EXERCISE 3. Let $\mathfrak{g}$ be a Lie algebra. Show that $\mathrm{ad}(X)$ is a derivation of $\mathfrak{g}$ for $X \in \mathfrak{g}$. Show also that $\mathrm{ad}(\mathfrak{g}) = \{\mathrm{ad}(X) | X \in \mathfrak{g}\}$ is an ideal of Lie algebra $\mathrm{Der}(\mathfrak{g})$ consisting of all derivations of $\mathfrak{g}$.

At this point we discuss several concepts related to representations. Let $\Phi\colon \mathfrak{g} \to \mathfrak{gl}(V)$ be a representation of a Lie algebra $\mathfrak{g}$ on V $(\neq \{0\})$. A subspace W of V is called a $\mathfrak{g}$*-invariant subspace* if W is stable under all $\Phi(X)$ $(X \in \mathfrak{g})$, i.e., $\Phi(X)(W) \subseteqq W$ for all $X \in \mathfrak{g}$. When W is $\mathfrak{g}$-invariant, the mapping $X \mapsto \Phi(X)|W$ (= the restriction of $\Phi(X)$ to W) is a representation of $\mathfrak{g}$ on W, which is denoted by Φ_W. The mapping which assigns to X the linear transformation of the factor space V/W induced by $\Phi(X)$ is also a representation of $\mathfrak{g}$ on V/W, which we denote by $\Phi_{V/W}$.

Given a representation on V, V and $\{0\}$ are clearly $\mathfrak{g}$-invariant subspaces. A representation on $V \neq \{0\}$ which has no $\mathfrak{g}$-invariant subspaces except V and $\{0\}$ is called *irreducible.* And if the representation is not irreducible, it is called *reducible.* If there exists a $\mathfrak{g}$-invariant subspace W'

such that $V = W \oplus W'$ (direct sum) for any $\mathfrak{g}$-invariant subspace W, then such a representation is called *completely reducible*. If a representation Φ on V is completely reducible, it is proved that there exist $\mathfrak{g}$-invariant subspaces $W_1, \ldots, W_r$ such that $V = W_1 \oplus \cdots \oplus W_r$ (direct sum) where the restrictions Φ_{W_i} are irreducible for all i.

EXERCISE 4. Verify these assertions.

EXAMPLE 4.7. For the adjoint representation of a Lie algebra $\mathfrak{g}$, the $\mathfrak{g}$-invariant subspaces are the ideals of $\mathfrak{g}$ and vice versa.

§2. Replicas of matrices

In order to prepare for the further discussion of Lie algebras, we introduce the theory([1]) of replicas of a matrix originated by Chevalley. We start with the following results on matrices. As before k denotes a field of characteristic 0.

THEOREM I. *Let A be an $n \times n$ matrix with coefficients in k. $Z(A)$ denotes the set of matrices which commute with A. Then a matrix X which commutes with all elements of $Z(A)$ can be expressed as a polynomial in A. Namely, there exists a polynomial $P(x) = \sum_{i=0}^n \alpha_i x^i$ $(\alpha_i \in k)$ such that $X = \sum_{i=0}^n \alpha_i A^i$.*

For a proof, see, e.g., N. Jacobson, *Lectures in abstract algebra*, vol. II, van Nostrand (republished by Springer), p. 113.

DEFINITION 4.3. An $n \times n$ matrix A is said to be *semisimple* if it is similar to a diagonal matrix over some extension field of k, and *nilpotent* if there exists an integer ν such that $A^\nu = 0$.

THEOREM II (Jordan decomposition). *An $n \times n$ matrix A can be written uniquely as a sum $A = S + N$, where S is semisimple, N is nilpotent, and $SN = NS$. Moreover, S and N can be expressed as polynomials in A without constant term. (That is, there are polynomials $P_1(x)$ and $P_2(x)$ such that $S = P_1(A)$, $N = P_2(A)$, and $P_1(0) = P_2(0) = 0$.) S and N are called the semisimple part and the nilpotent part of A respectively.*

For a proof, see, e.g., I. Satake, *Linear algebra*, Marcel Dekker, p. 168.([2])

Let V be an n-dimensional k-vector space. We can define semisimple and nilpotent linear transformations. And, for a linear transformation A, the semisimple part and nilpotent part of A are also defined and denoted by $A^{(s)}$ and $A^{(n)}$ respectively. They are the linear transformations whose

---

([1])C. Chevalley, *A new kind of relationship between matrices*, Amer. J. Math. **65** (1943), 521–531.

([2])To prove that we can take $P_1(x)$ with $P_1(0) = 0$, use the same argument as in Theorem 4.1 below. If A is regular, we can replace the constant term by means of the minimal polynomial of A. If not, use a common 0-eigenvector $v \neq 0$ of A and S, and prove $P_1(0) = 0$, $P_2(x) = x - P_1(x)$.

matrices are the semisimple part and nilpotent part of the matrix of A with respect to any basis.

For simplicity, we denote by $T_{p,q}$ the tensor space $T_q^{\,p}(V)$ of type (p, q). A linear transformation A of V defines a linear transformation $A_{p,q}$ of $T_{p,q}$ satisfying the following condition (4.1) (see §1.7)

$$\begin{aligned} A_{p,q}(v_1 \otimes v_2 \otimes \cdots \otimes v_p \otimes \varphi_1 \otimes \cdots \otimes \varphi_q) \\ = A(v_1) \otimes v_2 \otimes \cdots \otimes v_p \otimes \varphi_1 \otimes \cdots \otimes \varphi_q \\ + v_1 \otimes A(v_2) \otimes \cdots \otimes v_p \otimes \varphi_1 \otimes \cdots \otimes \varphi_q + \cdots \\ - v_1 \otimes \cdots \otimes v_p \otimes {}^tA(\varphi_1) \otimes \cdots \otimes \varphi_q - \cdots \\ - v_1 \otimes \cdots \otimes v_p \otimes \varphi_1 \otimes \cdots \otimes {}^tA(\varphi_q) \qquad (v_i \in V,\ \varphi_j \in V^*). \end{aligned} \tag{4.1}$$

If we make use of the Lie algebra $\mathfrak{gl}(V)$ of linear transformations of V, $A_{p,q}$ can be defined as follows: The mapping which assigns X to $X \in \mathfrak{gl}(V)$ is a representation of $\mathfrak{gl}(V)$ on V. The mapping $A \mapsto A_{p,q}$ is the tensor product of p copies of this representation and q copies of its contragredient representation. Therefore, this mapping is a Lie algebra homomorphism and we obtain equations (1) and (2) of the next lemma. (We can verify them directly without using the theory of Lie algebras.)

LEMMA 4.1. *Let $A, B \in \mathfrak{gl}(V)$ and $\alpha, \beta \in k$.*

(1) $(\alpha A + \beta B)_{p,q} = \alpha A_{p,q} + \beta B_{p,q}$.

(2) $[A, B]_{p,q} = [A_{p,q}, B_{p,q}]$.

(3) *If we identify naturally the tensor space $T_{q'}^{\,p'}(T_q^{\,p}(V))$ with $T_{pq'+qp'}^{\,pp'+qq'}(V)$,*

$$(A_{p,q})_{p',q'} = A_{pp'+qq',\,pq'+qp'}.$$

DEFINITION 4.4. Let $A \in \mathfrak{gl}(V)$. If $t \in T_{p,q}$ satisfies $A_{p,q}(t) = 0$, then t is called an *A-invariant tensor of type (p, q)*. The set of A-invariant tensors of type (p, q) is denoted by $I_{p,q}(A)$. $I_{p,q}(A)$ is obviously a subspace of $T_{p,q}$.

EXAMPLE 4.8.

$$I_{1,0}(A) = \{v \in V | A(v) = 0\},$$
$$I_{0,1}(A) = \{\varphi \in V^* | \varphi(A(v)) = 0 \text{ for all } v \in V\}.$$

LEMMA 4.2. *By the correspondence $\varphi \otimes v \leftrightarrow F_{\varphi,v}$, where $F_{\varphi,v}(v') = \varphi(v')v$, we can identify $T_{1,1}$ with $\mathfrak{gl}(V) = \mathrm{Hom}(V, V)$ (see Example 2.1, §2.1). A linear transformation $A_{1,1}$ of $T_{1,1}$ induces that of $\mathfrak{gl}(V)$ via this identification. We denote it by the same notation $A_{1,1}$. Then we have, for $X \in \mathfrak{gl}(V)$,*

$$A_{1,1}(X) = [A, X].$$

In particular, $X \in \mathfrak{gl}(V)$ corresponds an A-invariant tensor of type $(1, 1)$ if and only if $AX = XA$.

PROOF. By linearity of $A_{1,1}$, it suffices to prove the result for $X = F_{\varphi,v}$ ($\varphi \in V^*$, $v \in V$). To $A_{1,1}(F_{\varphi,v})$ there corresponds the following element of $T_{1,1}$:

$$A_{1,1}(\varphi \otimes v) = -{}^tA(\varphi) \otimes v + \varphi \otimes A(v).$$

We can compute the element of $\mathrm{Hom}(V, V)$ which corresponds to the right side. For all $v' \in V$,

$$\begin{aligned} F_{{}^tA(\varphi),v}(v') &= ({}^tA(\varphi))(v')v = \varphi(A(v'))v \\ &= F_{\varphi,v}(A(v')) = X(A(v')), \\ F_{\varphi,A(v)}(v') &= \varphi(v')A(v) = A(\varphi(v')v) \\ &= A(F_{\varphi,v}(v')) = A(X(v')). \end{aligned}$$

Thus, $A_{1,1}(X) = AX - XA = [A, X]$. □

EXAMPLE 4.9. Identify $T_{1,2} = V^* \otimes V^* \otimes V$ with $\mathscr{L}(V, V; V)$ (Example 2.4, §2.1). We also denote by $A_{1,2}$ the linear transformation of $\mathscr{L}(V, V; V)$ induced by $A_{1,2}$. Then, for $\Phi \in \mathscr{L}(V, V; V)$,

$$\begin{aligned} A_{1,2}(\Phi)(v_1, v_2) = A(\Phi(v_1, v_2)) &- \Phi(A(v_1), v_2) \\ &- \Phi(v_1, A(v_2)) \qquad (v_1, v_2 \in V). \end{aligned} \tag{4.2}$$

We shall prove this for Φ to which corresponds a tensor $\varphi_1 \otimes \varphi_2 \otimes v$. The general case follows.

$$\begin{aligned} A_{1,2}(\varphi_1 \otimes \varphi_2 \otimes v) = &-{}^tA(\varphi_1) \otimes \varphi_2 \otimes v - \varphi_1 \otimes {}^tA(\varphi_2) \otimes v \\ &+ \varphi_1 \otimes \varphi_2 \otimes A(v). \end{aligned}$$

Thus the value on $(v_1, v_2) \in V \times V$ of the corresponding element of $\mathscr{L}(V, V; V)$ is given by

$$\begin{aligned} &\varphi_1(v_1)\varphi_2(v_2)A(v) - \varphi_1(A(v_1))\varphi_2(v_2)v - \varphi_1(v_1)\varphi_2(A(v_2))v \\ &\quad = A(\varphi_1(v_1)\varphi_2(v_2)v) - \varphi_1(A(v_1))\varphi_2(v_2)v - \varphi_1(v_1)\varphi_2(A(v_2))v, \end{aligned}$$

which implies (4.2).

Let $(\mathfrak{A}, m)$ be an algebra and $V = \mathfrak{A}$. From the above correspondence $\mathscr{L}(\mathfrak{A}, \mathfrak{A}; \mathfrak{A}) \cong \mathfrak{A}^* \otimes \mathfrak{A}^* \otimes \mathfrak{A}$, to the multiplication m there corresponds a tensor t_m. Then, by (4.2), $D \in \mathfrak{gl}(\mathfrak{A})$ is a derivation if and only if $D_{1,2}(t_m) = 0$.

EXERCISE. $T_{0,2} = V^* \otimes V^*$ can be identified with $\mathscr{L}(V, V; k)$ (Example 2.2., §2.1). We also denote by $A_{0,2}$ the linear transformation of $\mathscr{L}(V, V; k)$ induced by $A_{0,2}$. Show that, for $\Phi = \mathscr{L}(V, V; k)$,

$$A_{0,2}(\Phi)(v_1, v_2) = -\Phi(A(v_1), v_2) - \Phi(v_1, A(v_2)) \qquad (v_1, v_2 \in V).$$

Thus, Φ corresponds to an element of $I_{0,2}(A)$ if and only if

$$\Phi(A(v_1), v_2) + \Phi(v_1, A(v_2)) = 0$$

for all $v_1, v_2 \in V$.

DEFINITION 4.5 (Chevalley). Let A and B be elements of $\mathfrak{gl}(V)$. When $I_{p,q}(A) \subseteqq I_{p,q}(B)$ for all p and q, B is called a *replica* of A. To denote that B is a replica of A, we write $A \twoheadrightarrow B$.

LEMMA 4.3. *Let $A, B, C \in \mathfrak{gl}(V)$.*

(1) $A \twoheadrightarrow A$.

(2) If $A \twoheadrightarrow B$ and $B \twoheadrightarrow C$, then $A \twoheadrightarrow C$.

(3) If $A \twoheadrightarrow B$, then $A_{p,q} \twoheadrightarrow B_{p,q}$.

(4) If $A \twoheadrightarrow B$ and $A \twoheadrightarrow C$, then $A \twoheadrightarrow \alpha B + \beta C$ $(\alpha, \beta \in k)$. In particular, the set of replicas of A is a subspace of $\mathfrak{gl}(V)$.

PROOF. (1), (2), and (4) follow from the definition. (3) is obtained from Lemma 4.1 (3).

THEOREM 4.1. *Let $A, B \in \mathfrak{gl}(V)$. If $A \twoheadrightarrow B$, then B can be expressed as a polynomial in A without constant term. Namely, there exists a polynomial $P(x)$ in x such that $B = P(A)$ and $P(0) = 0$.*

PROOF. By Lemma 4.2, if $X \in \mathfrak{gl}(V)$ satisfies $AX = XA$, then X corresponds to an element of $I_{1,1}(A)$. Since $I_{1,1}(A) \subseteqq I_{1,1}(B)$, X commutes with B. Thus, B can be expressed as polynomial $P(A)$ in A by Theorem I.

Let us show that we can choose P such that $P(0) = 0$. If A is regular, then 0 is not a root of the minimal polynomial of A. Thus, a constant can be expressed as a polynomial in A without constant term. We can substitute it in P if P has nonzero constant term. If A is not regular, it follows that $P(0) = 0$. In fact, there exists a nonzero $v \in V = T_{1,0}$ such that $A(v) = 0$. Since $v \in I_{1,0}(A) \subseteqq I_{1,0}(B)$, $B(v) = 0$, i.e., $P(A)(v) = 0$. $A(v) = 0$ implies $P(0)(v) = 0$. As $v \neq 0$, we have $P(0) = 0$. □

COROLLARY 1. *$A \twoheadrightarrow B$ if and only if $B_{p,q}$ can be expressed as a polynomial in $A_{p,q}$ without constant term for all (p, q).*

PROOF. If $A \twoheadrightarrow B$, then $A_{p,q} \twoheadrightarrow B_{p,q}$ by Lemma 4.3(3). Thus, by the theorem, $B_{p,q}$ is a polynomial in $A_{p,q}$ without constant term. The converse clearly follows from the definition of replica. □

COROLLARY 2. *Let $W_1 \supseteqq W_2$ be subspaces of $T_{p,q}$ such that $A_{p,q}(W_1) \subseteqq W_2$. If $A \twoheadrightarrow B$, then $B_{p,q}(W_1) \subseteqq W_2$.*

PROOF. Since $B_{p,q}$ is a polynomial in $A_{p,q}$ without constant term, the result follows from $A_{p,q}(W_1) \subseteqq W_2$. □

PROPOSITION 4.1. *Let A, B be elements of $\mathfrak{gl}(V)$ such that $A \twoheadrightarrow B$. If k_1 is an extension field of k, we have $A^{k_1} \twoheadrightarrow B^{k_1}$. Conversely, a replica of A^{k_1} can be expressed as a linear combination of replicas of A with coefficients in k_1. (For the definition of A^{k_1}, see Proposition 1.13.)*

PROOF. By Corollary 1 to Theorem 4.1, $B_{p,q}$ can be expressed as a polynomial $P_{p,q}(A_{p,q})$ in $A_{p,q}$ with coefficients in k. As we can see easily, $(B_{p,q})^{k_1} = (P_{p,q}(A_{p,q}))^{k_1} = P_{p,q}((A_{p,q})^{k_1}) = P_{p,q}((A^{k_1})_{p,q})$, which implies $A^{k_1} \twoheadrightarrow B^{k_1}$ by Corollary 1 again.

Let us prove the converse. In general, a k_1-linear transformation $\widetilde{X}$ of V^{k_1} can be expressed in the form of $\widetilde{X} = \sum_\lambda \omega_\lambda X_\lambda^{k_1}$, where the ω_λ are elements of k_1 which are linearly independent over k and the X_λ are linear transformations of V. (In fact, take a basis $(e_1, \dots, e_n)$ of V and construct a basis $(1\otimes e_1, \dots, 1\otimes e_n)$ of V^{k_1}. Suppose that $\widetilde{X}(1\otimes e_i) = \sum_{j=1}^n \xi_{ji}\otimes e_j$ $(\xi_{ji} \in k_1)$. ξ_{ji} can be written in the form $\xi_{ji} = \sum_\lambda \omega_\lambda \xi_{ji}{}^\lambda$ $(\xi_{ji}{}^\lambda \in k)$. Define a linear transformation X_λ of V by $X_\lambda(e_i) = \sum_j \xi_{ji}{}^\lambda e_j$ $(j = 1, \dots, n)$. Then, we have $\widetilde{X} = \sum_\lambda \omega_\lambda X_\lambda^{k_1}$.) We know that $\widetilde{X}_{p,q} = \sum \omega_\lambda (X_\lambda)_{p,q}{}^{k_1}$. Choose $t \in I_{p,q}(\widetilde{X}) \cap T_q{}^p(V)$. (We can naturally consider that $T_q{}^p(V) \subseteqq T_q{}^p(V^{k_1})$.) Then $\widetilde{X}_{p,q}(t) = \sum \omega_\lambda (X_\lambda)_{p,q}(t) = 0$, which implies that $(X_\lambda)_{p,q}(t) = 0$ for all λ, since the ω_λ are linearly independent.

Suppose that $A^{k_1} \twoheadrightarrow \widetilde{X}$ and write $\widetilde{X} = \sum \omega_\lambda X_\lambda{}^{k_1}$ as above. We shall prove that $A \twoheadrightarrow X_\lambda$. Take $t \in I_{p,q}(A)$. When $t \in I_{p,q}(A) \subseteqq T_q{}^p(V)$ is regarded as an element of $T_q{}^p(V^{k_1})$, $t \in I_{p,q}(A^{k_1})$ and $t \in I_{p,q}(\widetilde{X})$ because of $A^{k_1} \twoheadrightarrow \widetilde{X}$. From the above remark, $t \in I_{p,q}(X_\lambda)$, which implies $I_{p,q}(A) \subseteqq I_{p,q}(X_\lambda)$ and hence $A \twoheadrightarrow X_\lambda$. □

In particular, for linear transformations A, B of V, $A \twoheadrightarrow B$ is equivalent to $A^{k_1} \twoheadrightarrow B^{k_1}$.

Next, we discuss the relation between Jordan decompositions and replicas.

LEMMA 4.4. *If $A \in \mathfrak{gl}(V)$ is semisimple (resp. nilpotent), then $A_{p,q}$ is semisimple (resp. nilpotent).*

PROOF. Suppose that A is semisimple. $-{}^tA$ is clearly semisimple and tensor products of them with the identity transformation are also semisimple. (We can see this from the fact that, if a linear transformation is semisimple, we can take a basis consisting of eigenvectors after extending the field.) Thus $A_{p,q}$ is a sum of semisimple linear transformations which commute each other. Therefore $A_{p,q}$ is semisimple.

When A is nilpotent, we can argue similarly. □

PROPOSITION 4.2. (1) *Let $A^{(\mathrm{s})}$ and $A^{(\mathrm{n})}$ be the semisimple part and the nilpotent part of A respectively. Then, for all (p, q),*

$$(A_{p,q})^{(\mathrm{s})} = (A^{(\mathrm{s})})_{p,q}, \qquad (A_{p,q})^{(\mathrm{n})} = (A^{(\mathrm{n})})_{p,q}.$$

(2) *Therefore, $A \twoheadrightarrow A^{(\mathrm{s})}$ and $A \twoheadrightarrow A^{(\mathrm{n})}$.*

PROOF. (1) By Lemma 4.4, $(A^{(s)})_{p,q}$ and $(A^{(n)})_{p,q}$ are semisimple and nilpotent respectively. Moreover, they commute and we have

$$A_{p,q} = (A^{(s)} + A^{(n)})_{p,q} = (A^{(s)})_{p,q} + (A^{(n)})_{p,q}.$$

From the uniqueness of Jordan decompositions, we have the result.

(2) By Theorem II, $(A^{(s)})_{p,q} = (A_{p,q})^{(s)}$ is a polynomial in $A_{p,q}$ without constant term. Therefore, by Corollary 1 to Theorem 4.1, we have $A \twoheadrightarrow A^{(s)}$. Similarly $A \twoheadrightarrow A^{(n)}$. □

PROPOSITION 4.3. *Let A, B be elements of $\mathfrak{gl}(V)$ such that $A \twoheadrightarrow B$.*

(1) *If A is semisimple (resp. nilpotent), then B is also semisimple (resp. nilpotent).*

(2) *If B is semisimple, then we have $A^{(s)} \twoheadrightarrow B$. If B is nilpotent, then $A^{(n)} \twoheadrightarrow B$.*

PROOF. Since B is a polynomial in A without constant term, (1) obviously holds.

(2) Suppose that B is semisimple. Choose an element $t \in I_{p,q}(A^{(s)})$. Let W be the subspace of $T_{q,p}$ generated by $\{t, (A_{p,q})(t), (A_{p,q})^2(t), \dots\}$. By definition, we have $A_{p,q}(W) \subseteqq W$. Therefore, by Corollary 2 to Theorem 4.1, $B_{p,q}(W) \subseteqq W$. We denote by $(A_{p,q})_W$, $(B_{p,q})_W$, ... the restriction of $A_{p,q}$, $B_{p,q}$, $A^{(s)}_{p,q}$, $A^{(n)}_{p,q}$ to W. Since $A^{(s)}_{p,q}(W) = \{0\}$, $(A_{p,q})_W = (A^{(s)}_{p,q})_W + (A^{(n)}_{p,q})_W = (A^{(n)}_{p,q})_W$, which shows that $(A_{p,q})_W$ is nilpotent. $(B_{p,q})_W$ is also nilpotent, since it is a polynomial in $(A_{p,q})_W$ without constant term. On the other hand, since B is semisimple, $(B_{p,q})_W$ is semisimple. Being both nilpotent and semisimple, $(B_{p,q})_W$ is equal to 0. In particular, $B_{p,q}(t) = 0$, which proves $A^{(s)} \twoheadrightarrow B$.

When B is nilpotent, we can prove $A^{(n)} \twoheadrightarrow B$ in a similar fashion. □

THEOREM 4.2. *Let $A, B \in \mathfrak{gl}(V)$. $A \twoheadrightarrow B$ if and only if $A^{(s)} \twoheadrightarrow B^{(s)}$ and $A^{(n)} \twoheadrightarrow B^{(n)}$ hold.*

PROOF. Suppose that $A \twoheadrightarrow B$. Since $B \twoheadrightarrow B^{(s)}$, we have $A \twoheadrightarrow B^{(s)}$. From Proposition 4.3(2), $A^{(s)} \twoheadrightarrow B^{(s)}$. Similarly, we have $A^{(n)} \twoheadrightarrow B^{(n)}$.

Conversely, suppose that $A^{(s)} \twoheadrightarrow B^{(s)}$ and $A^{(n)} \twoheadrightarrow B^{(n)}$. Since $A \twoheadrightarrow A^{(s)}$ and $A \twoheadrightarrow A^{(n)}$, we have $A \twoheadrightarrow B^{(s)} + B^{(n)} = B$. □

This theorem shows that, in order to determine the replicas of a linear transformation, it is enough to consider the replicas of its semisimple and nilpotent parts. In fact, let $A \in \mathfrak{gl}(V)$ and suppose that $A^{(s)} \twoheadrightarrow B_1$ and $A^{(n)} \twoheadrightarrow B_2$. Then B_1, B_2 are semisimple and nilpotent respectively. Moreover, since B_1 and B_2 are polynomials in $A^{(s)}$ and $A^{(n)}$ respectively, they are polynomials in A and commute. Thus B_1 and B_2 are the semisimple and nilpotent parts of $B = B_1 + B_2$ respectively. By Theorem 4.2, we have $A \twoheadrightarrow B$. Now we shall determine the replicas of nilpotent and semisimple transformations.

THEOREM 4.3. *Let $A \in \mathfrak{gl}(V)$ be nilpotent. Then $A \twoheadrightarrow B$ if and only if $B = \alpha A \quad (\alpha \in k)$.*([3])

PROOF. If $B = \alpha A$, we clearly have $A \twoheadrightarrow B$. Let us prove the converse. Since A is nilpotent, there exists a natural number ρ which satisfies $A^{\rho-1} \neq 0$ and $A^{\rho} = 0$. Then $\{A^0 \ (= I), A, \dots, A^{\rho-1}\} \subseteqq \mathfrak{gl}(V)$ is linearly independent and $\{A^i \otimes A^j | 0 \le i, j \le \rho - 1\}$ is a linearly independent subset of $\mathfrak{gl}(V \otimes V)$.

By $A_{2,0} = A \otimes I + I \otimes A$ and $A^{\rho} = 0$, we have

$$(A_{2,0})^l = \sum_{i=0}^{l} \binom{l}{i} A^i \otimes A^{l-i} \tag{4.3}$$
$$= \begin{cases} \sum_{i=0}^{l} \binom{l}{i} A^i \otimes A^{l-i} & (\text{if } l < \rho), \\ \sum_{i=l-\rho+1}^{\rho-1} \binom{l}{i} A^i \otimes A^{l-i} & (\text{if } \rho \le l \le 2\rho - 2), \\ 0 & (\text{if } l \ge 2\rho - 1). \end{cases}$$

Suppose that $A \twoheadrightarrow B$. Then $A_{2,0} \twoheadrightarrow B_{2,0}$ and, by Theorem 4.1, we can write:

$$B = \alpha_1 A + \alpha_2 A^2 + \cdots + \alpha_{\rho-1} A^{\rho-1} \qquad (\alpha_i \in k), \tag{4.4}$$

$$B_{2,0} = \beta_1 A_{2,0} + \beta_2 (A_{2,0})^2 + \cdots + \beta_{2\rho-2} (A_{2,0})^{2\rho-2} \qquad (\beta_j \in k). \tag{4.5}$$

By (4.4),

$$B_{2,0} = B \otimes I + I \otimes B = \sum_{i=1}^{\rho-1} \alpha_i (A^i \otimes I + I \otimes A^i).$$

On the other hand, if we substitute (4.3) in (4.5),

$$B_{2,0} = \sum_{l=1}^{\rho-1} \beta_l \left(\sum_{i=1}^{l} \binom{l}{i} A^i \otimes A^{l-i} \right) + \sum_{l=\rho}^{2\rho-2} \beta_l \left(\sum_{i=l-\rho+1}^{\rho-1} \binom{l}{i} A^i \otimes A^{l-i} \right).$$

Since $\{A^i \otimes A^j | 0 \le i, j \le \rho - 1\}$ is linearly independent, we compare these two identities and obtain that

$$\begin{aligned} \alpha_l &= \beta_l & (l = 1, \dots, \rho - 1), \\ \beta_l &= 0 & (l = \rho, \dots, 2\rho - 2). \end{aligned}$$

Moreover, if $\beta_l \neq 0 \ (l \ge 2)$ we must have $\binom{l}{1} = \cdots = \binom{l}{l-1} = 0$. But, since $\binom{l}{1} = l$, $\beta_l = \alpha_l = 0$ for $l \ge 2$, which implies $B = \alpha_1 A$. □

REMARK. When the field has characteristic p, the conclusion of the theorem above is:

$$B = \alpha_0 A + \alpha_1 A^p + \cdots + \alpha_f A^{p^f}.$$

([3])H. Tuan, *A note on the replicas of nilpotent matrices* Bull. Amer. Math. Soc. **51** (1945), 305–312 and I. S. Cohen, *Note on a note of H. F. Tuan* Bull. Amer. Math. Soc. **52** (1946), 175–177.

To prove this, after reasoning in the same way, we use the following result; If $l \geq 2$, then the greatest common divisor of $\binom{l}{1}, \ldots, \binom{l}{l-1}$ is 1, when l has more than 2 prime factors. And the greatest common divisor is equal to p, when l is a power of a prime number p.

Next, we shall determine the replicas of a semisimple transformation.

THEOREM 4.4. *Let A be a semisimple transformation, let $\alpha_1, \ldots, \alpha_n$ be eigenvalues of A, and let $P(x)$ be a polynomial in x over the field k. Then $A \twoheadrightarrow P(A)$ if and only if the following condition is satisfied: For all linear relations, $\sum_i n_i\alpha_i = 0$ $(n_i \in \mathbf{Z})$ between $\alpha_1, \ldots, \alpha_n$ with integral coefficients, we have $\sum_i n_i P(\alpha_i) = 0$.*

PROOF. Considering some extension field of k, if necessary, we may assume that the matrix of A is a diagonal matrix, in other words, that there exists a basis $(v_1, \ldots, v_n)$ of V consisting of eigenvectors of A $(Av_i = \alpha_i v_i)$. Then the matrix of $A_{p,q}$ with respect to the standard basis of $T_{p,q} = T_q^{\,p}(V)$ corresponding to this basis is also a diagonal matrix, whose diagonal components are all the elements of k of the form

$$\sum_{\lambda=1}^{p} \alpha_{i_\lambda} - \sum_{\mu=1}^{q} \alpha_{j_\mu} \qquad (1 \leq i_\lambda \leq n,\ 1 \leq j_\mu \leq n).$$

In fact, for a basis element $t_{i_1,\ldots,i_p}^{\;j_1,\ldots,j_q}$ of $T_{p,q}$, we have

$$A_{p,q}(t_{i_1,\ldots,i_p}^{\;j_1,\ldots,j_q}) = (\alpha_{i_1} + \cdots + \alpha_{i_p} - \alpha_{j_1} - \cdots - \alpha_{j_q})t_{i_1,\ldots,i_p}^{\;j_1,\ldots,j_q}.$$

(For the dual basis $(\varphi_1, \ldots, \varphi_n)$ of $(v_1, \ldots, v_n)$, $({}^tA)(\varphi_i) = \alpha_i\varphi_i$.)

The condition $A \twoheadrightarrow P(A)$ is, by definition, for all $t \in T_{p,q}$, such that $A_{p,q}(t) = 0$, $P(A)_{p,q}(t) = 0$, which can be rephrased as saying that the eigenspace of $A_{p,q}$ for the eigenvalue 0 is contained in the eigenspace of $P(A)_{p,q}$ for the eigenvalue 0. Therefore, as we can see by computing the matrix of $P(A)$ with respect to the above basis, this condition is equivalent to the condition that, if $\sum_{\lambda=1}^{p} \alpha_{i_\lambda} - \sum_{\mu=1}^{q} \alpha_{j_\mu} = 0$, $\sum_{\lambda=1}^{p} P(\alpha_{i_\lambda}) - \sum_{\mu=1}^{q} P(\alpha_{j_\mu}) = 0$. The validity of this condition for all p and q is the assertion of the theorem. □

EXAMPLE 4.10. Let $k = \mathbf{C}$. Assume that, with respect to a basis of V, the matrix of $A \in \mathfrak{gl}(V)$ is given by

$$\begin{bmatrix} \alpha_1 & 0 & 0 \\ 0 & \alpha_2 & 0 \\ 0 & 0 & \alpha_3 \end{bmatrix}.$$

Let us study the polynomials $P(x)$ such that $A \twoheadrightarrow P(A)$. The degree d of the minimal polynomial of A is equal to the dimension of the subspace of $\mathfrak{gl}(V)$ generated by $\{A^m \mid m = 0, 1, \ldots\}$. Since we can take $(I, A, \ldots,$

A^{d-1}) as basis of this subspace, we may assume that the degree of $P(x)$ is less than or equal to $d-1$. We know that d is equal to the number of different elements of $\alpha_1, \alpha_2, \alpha_3$. We denote by Replica($A$) the subspace of replicas of A.

When $d = 1$, A is a scalar matrix. Polynomials in A without constant term are scalar multiples of A that are replicas of A.

Let Λ be the subspace generated by $\{\alpha_1, \alpha_2, \alpha_3\}$ over $\mathbf{Q}$ (the field of rational numbers). Let us consider the following three cases separately, depending on the dimension of Λ.

(1) *The case* $\dim \Lambda = 3$ (namely, $\alpha_1, \alpha_2, \alpha_3$ are linearly independent over $\mathbf{Q}$.) In this case, $d = 3$. The relation $\sum n_i \alpha_i = 0$ $(n_i \in \mathbf{Z})$ among α_i implies $n_i = 0$ for all i. Therefore $A \twoheadrightarrow P(A)$ for all polynomials $P(x)$ and Replica(A) is of dimension 3. A replica of A can be expressed in the form $\xi_0 I + \xi_1 A + \xi_2 A^2$ $(\xi_0, \xi_1, \xi_2 \in \mathbf{C})$. A replica of A can be also expressed in the form $\eta_1 A + \eta_2 A^2 + \eta_3 A^3$ $(\eta_1, \eta_2, \eta_3 \in \mathbf{C})$, since I is a polynomial in A.

(2) *The case* $\dim \Lambda = 2$. Changing the order, if necessary, we may assume that α_1 and α_2 are linearly independent over $\mathbf{Q}$, and $\alpha_3 = (m_1/n_0)\alpha_1 + (m_2/n_0)\alpha_2$, where $n_0, m_1, m_2 \in \mathbf{Z}$, $n_0 \neq 0$. Since α_1 and α_2 are linearly independent over $\mathbf{Q}$, we can conclude that the relations with coefficients in $\mathbf{Z}$ among $\alpha_1, \alpha_2, \alpha_3$ are scalar multiples of $n_0\alpha_3 - m_1\alpha_1 - m_2\alpha_2 = 0$. Therefore, by Theorem 4.4, for a polynomial $P(x)$, $A \twoheadrightarrow P(A)$ if and only if $P(x)$ satisfies

$$n_0 P(\alpha_3) - m_1 P(\alpha_1) - m_2 P(\alpha_2) = 0.$$

(i) $d = 2$. Since α_3 is equal to α_1 or α_2 $(\alpha_1 \neq \alpha_2)$, and the relations are trivial ones deduced from $\alpha_3 = \alpha_1$ or $\alpha_3 = \alpha_2$, we have $A \twoheadrightarrow P(A)$ for all polynomials and Replica(A) is of dimension 2.

(ii) $d = 3$. Set $P(x) = \xi_0 + \xi_1 x + \xi_2 x^2$ and substitute this into the above equality. Then

$$\xi_0(n_0 - m_1 - m_2) + \xi_2(n_0\alpha_3^2 - m_1\alpha_1^2 - m_2\alpha_2^2) = 0.$$

Since $\alpha_1, \alpha_2, \alpha_3$ are all different, the coefficients of ξ_0 and ξ_2, $n_0 - m_1 - m_2$ and $n_0\alpha_3^2 - m_1\alpha_1^2 - m_2\alpha_2^2$, cannot be simultaneously equal to 0. Therefore Replica(A) is of dimension 2.

(3) *The case* $\dim \Lambda = 1$. Suppose that $\alpha_1 \neq 0$, $\alpha_2 = (m_2/n_0)\alpha_1$, and $\alpha_3 = (m_3/n_0)\alpha_1$, where $n_0, m_2, m_3 \in \mathbf{Z}$, $n_0 \neq 0$. Similarly to the case (2), $A \twoheadrightarrow P(A)$ for a polynomial $P(x)$ if and only if we have the following two equalities,

$$n_0 P(\alpha_2) - m_2 P(\alpha_1) = 0, \qquad n_0 P(\alpha_3) - m_3 P(\alpha_1) = 0.$$

Set

$$P(x) = \begin{cases} \xi_0 + \xi_1 x & \text{if } d = 2, \\ \eta_0 + \eta_1 x + \eta_2 x^2 & \text{if } d = 3, \end{cases}$$

and substitute it into the above equalities. Then we obtain $\xi_0 = 0$ and $\eta_0 = \eta_2 = 0$ respectively. This implies that Replica(A) is of dimension 1 and consists of the scalar multiples of A.

The next theorem is a significant application of the theory of replicas.

THEOREM 4.5. *Let* $A \in \mathfrak{gl}(V)$. *A is nilpotent if and only if* $\operatorname{tr}(AB) = 0$ *for all replicas B of A.*

PROOF. If A is nilpotent and $A \twoheadrightarrow B$, then AB is nilpotent and $\operatorname{tr}(AB) = 0$.

Conversely, let us show that, if $\operatorname{tr}(AB) = 0$ for all replicas B, then A is nilpotent. Let $A = A^{(s)} + A^{(n)}$ be the Jordan decomposition. It suffices to show that $A^{(s)} = 0$. In order to do this, we shall prove the following assertion.

(∗) Let S be a semisimple transformation. If $\operatorname{tr}(SB) = 0$ for all replicas B of S, then $S = 0$.

Suppose we have proved (∗) and that $\operatorname{tr}(AB) = 0$ for all replicas B of A. Let B_s be a replica of $S = A^{(s)}$. Then $A \twoheadrightarrow B_s$. Since B_s commutes with $A^{(n)}$, $A^{(n)}B_s$ is nilpotent and $\operatorname{tr}(A^{(n)}B_s) = 0$. On the other hand, since $A \twoheadrightarrow B_s$, $\operatorname{tr}(AB_s) = 0$ by assumption. Therefore $\operatorname{tr}(A^{(s)}B_s) = \operatorname{tr}(AB_s) - \operatorname{tr}(A^{(n)}B_s) = 0$. From (∗), we have $A^{(s)} = 0$, which proves the theorem.

Now, let us prove (∗). We may assume, in some extension field of k if necessary, that S can be expressed by a diagonal matrix. Let $\alpha_1, \dots, \alpha_n$ be the eigenvalues of S and Γ the k_0-vector space generated by $\alpha_1, \dots, \alpha_n$, where k_0 is the prime field (isomorphic to $\mathbf{Q}$). To prove $S = 0$, i.e., $\alpha_1 = \cdots = \alpha_n = 0$, it suffices to show that $\Gamma = \{0\}$, which is equivalent to $\Gamma^* = \{0\}$. Let φ be an element of Γ^* and set $\varphi(\alpha_i) = \beta_i$. There exists a polynomial $P(x)$ such that $P(\alpha_i) = \beta_i$ for all i. Let $B = P(S)$ and $\sum n_i\alpha_i = 0$ be a linear relation with integer coefficients among the α_i. Applying the k_0-linear mapping φ, we have $0 = \sum n_i\varphi(\alpha_i) = \sum n_i\beta_i = \sum n_iP(\alpha_i)$. By Theorem 4.4, $S \twoheadrightarrow B$. Therefore, by assumption,

$$0 = \operatorname{tr}(SB) = \sum_i \alpha_i\beta_i = \sum \alpha_i\varphi(\alpha_i).$$

Applying φ again, where $\varphi(\alpha_i) \in k_0$, we have $0 = \sum_i(\varphi(\alpha_i))^2$, which implies $\varphi(\alpha_i) = 0$, that is, $\varphi = 0$. □

REMARK. To prove this theorem, it is necessary to assume that k has characteristic 0.

§3. Properties of Lie algebras

(a) Definition of nilpotent Lie algebras and solvable Lie algebras. Now, let us come back to the discussion of Lie algebras and define several important families of Lie algebras. Let $\mathfrak{g}$ be a Lie algebra and $\mathfrak{a}_1$, $\mathfrak{a}_2$ ideals of $\mathfrak{g}$.

Denote by $[\mathfrak{a}_1, \mathfrak{a}_2]$ the subspace generated by the elements $[X, Y]$, $X \in \mathfrak{a}_1$, $Y \in \mathfrak{a}_2$, namely,

$$[\mathfrak{a}_1, \mathfrak{a}_2] = \left\{ \sum_i [X_i, Y_i] | X_i \in \mathfrak{a}_1, \ Y_i \in \mathfrak{a}_2 \right\}.$$

Then $[\mathfrak{a}_1, \mathfrak{a}_2]$ is an ideal of $\mathfrak{g}$, because of the identity $[Z, [X, Y]] = [[Z, X], Y] + [X, [Z, Y]]$, which is obtained from the Jacobi identity.

For an integer $m \geq 0$, we define the ideals $\mathfrak{g}^{(m)}$ and $\mathfrak{g}_m$ inductively by the following formulas:

$$\mathfrak{g}^{(0)} = \mathfrak{g}_0 = \mathfrak{g},$$
$$\mathfrak{g}^{(m)} = [\mathfrak{g}^{(m-1)}, \mathfrak{g}^{(m-1)}], \quad \mathfrak{g}_m = [\mathfrak{g}, \mathfrak{g}_{m-1}] \quad (m \geq 1).$$

It follows clearly $\mathfrak{g}_m \supseteqq \mathfrak{g}^{(m)}$.

DEFINITION 4.6. (1) If $\mathfrak{g}^{(1)} = \mathfrak{g}_{(1)} = \{0\}$, then $\mathfrak{g}$ is called *commutative* or abelian).

(2) If there is an integer $m \geq 0$ such that $\mathfrak{g}_m = \{0\}$, then $\mathfrak{g}$ is called *nilpotent.*

(3) If there exists an integer $m \geq 0$ such that $\mathfrak{g}^{(m)} = \{0\}$, then $\mathfrak{g}$ is called *solvable.*

EXAMPLE 4.11. Let $\mathfrak{g}$ be a k-vector space. Define a multiplication $[\ ,\ ]$: $\mathfrak{g} \times \mathfrak{g} \to \mathfrak{g}$ to be the zero mapping $(X, Y) \mapsto 0$, then $\mathfrak{g}$ satisfies the conditions of Lie algebras (Definition 4.1(2)) and is a commutative Lie algebra.

From condition (2)(i) of Definition 4.1, Lie algebras of dimension 1 are always commutative.

EXAMPLE 4.12. Let $\mathfrak{g}$ be a noncommutative Lie algebra of dimension 2. Then $\mathfrak{g}^{(1)} = \mathfrak{g}_1 = [\mathfrak{g}, \mathfrak{g}]$ is of dimension 1 and (since $\mathfrak{g}^{(2)} = \{0\}$) $\mathfrak{g}$ is solvable. Let X_1 be a generator of $[\mathfrak{g}, \mathfrak{g}]$ and (X_1, X_2) a basis of $\mathfrak{g}$ containing X_1. By taking a suitable scalar multiple of X_2, if necessary, we may assume that $[X_1, X_2] = X_1$. Then, by definition of Lie algebras, the products of basis elements are given by

$$[X_1, X_1] = [X_2, X_2] = 0, \qquad [X_2, X_1] = -[X_1, X_2] = -X_1.$$

Conversely, let $\mathfrak{g}$ be a 2-dimensional vector space with a basis (X_1, X_2). Define a bilinear multiplication on $\mathfrak{g}$ as above. Then $\mathfrak{g}$ is a Lie algebra. (Consider the Lie algebra $\mathfrak{g} = \{\left[\begin{smallmatrix} a & b \\ 0 & -a \end{smallmatrix}\right] | a, b \in k\} \subseteqq \mathfrak{gl}_2(k)$ (cf. Example 4.2, §4.1). $\mathfrak{g}$ is not commutative and has dimension 2. This also shows the existence of such a Lie algebra.) Thus we know that there exists only one 2-dimensional Lie algebra $\mathfrak{g}$ up to isomorphism. For $\mathfrak{g}$, $\mathfrak{g}_m = \mathfrak{g}_1$ $(m \geq 2)$ and $\mathfrak{g}$ is not nilpotent.

EXAMPLE 4.13. Let $\mathfrak{b}$ and $\mathfrak{u}$ be the set of $n \times n$ upper triangular matrices (a_{ij}), $(a_{ij} = 0$ if $i > j)$ and the set of $n \times n$ upper triangular matrices whose diagonal components are all zero $(a_{ij} = 0$ if $i \geq j)$ respectively. $\mathfrak{b}$

and $\mathfrak{u}$ are Lie algebras as Lie subalgebras of $\mathfrak{gl}_n(k)$. We can verify by matrix computation that $\mathfrak{u}$ is nilpotent and $\mathfrak{b}$ is solvable.

EXAMPLE 4.14. By definition, Lie subalgebras and quotient algebras of nilpotent Lie algebras are nilpotent. A similar result holds for solvable Lie algebras.

PROPOSITION 4.4. *The following three conditions for a Lie algebra* $\mathfrak{g}$ *are equivalent*:

(1) $\mathfrak{g}$ *is a nilpotent Lie algebra.*

(2) *There exists a descending chain of ideals*

$$\mathfrak{g} = \mathfrak{a}_0 \supseteqq \mathfrak{a}_1 \supseteqq \cdots \supseteqq \mathfrak{a}_m = \{0\}$$

such that $[\mathfrak{g}, \mathfrak{a}_i] \subseteqq \mathfrak{a}_{i+1}$ *for all* i $(0 \le i \le m-1)$.

(3) *Let* $\mathfrak{z}_1$ *be the center of* $\mathfrak{g}$, $\mathfrak{z}_2/\mathfrak{z}_1$ *the center of* $\mathfrak{g}/\mathfrak{z}_1$. *In general, denote by* $\mathfrak{z}_i/\mathfrak{z}_{i-1}$ *the center of* $\mathfrak{g}/\mathfrak{z}_{i-1}$. *Then there exists an integer* m *such that* $\mathfrak{g} = \mathfrak{z}_m$.

PROOF. (1) $\Rightarrow$ (2). Set $\mathfrak{a}_i = \mathfrak{g}_i$. Then $\mathfrak{a}_i$ $(i = 1, \ldots)$ satisfy the condition.

(2) $\Rightarrow$ (3). Since $[\mathfrak{g}, \mathfrak{a}_{m-1}] = \{0\}$, $\mathfrak{a}_{m-1} \subseteqq \mathfrak{z}_1$. And $[\mathfrak{g}, \mathfrak{a}_{m-2}] \subseteqq \mathfrak{a}_{m-1} \subseteqq \mathfrak{z}_1$ shows that $\mathfrak{a}_{m-2} \subseteqq \mathfrak{z}_2$. Similarly we have $\mathfrak{a}_{m-i} \subseteqq \mathfrak{z}_i$ and $\mathfrak{g} = \mathfrak{a}_0 \subseteqq \mathfrak{z}_m$.

(3) $\Rightarrow$ (1). If $\mathfrak{g} = \mathfrak{z}_m$, then we have $\mathfrak{g}_1 = [\mathfrak{g}, \mathfrak{g}] \subseteqq \mathfrak{z}_{m-1}$, $\mathfrak{g}_2 = [\mathfrak{g}, \mathfrak{g}_1] \subseteqq [\mathfrak{g}, \mathfrak{z}_{m-1}] \subseteqq \mathfrak{z}_{m-2}$. Similarly $\mathfrak{g}_i \subseteqq \mathfrak{z}_{m-i}$, which shows $\mathfrak{g}_{m-1} \subseteqq \mathfrak{z}_1$ and $\mathfrak{g}_m = [\mathfrak{g}, \mathfrak{g}_{m-1}] \subseteqq [\mathfrak{g}, \mathfrak{z}_1] = \{0\}$. □

For a solvable Lie algebra, we have a similar proposition.

PROPOSITION 4.5. *The following conditions* (1) *and* (2) *for a Lie algebra* $\mathfrak{g}$ *are equivalent*:

(1) $\mathfrak{g}$ *is a solvable Lie algebra.*

(2) *There exists a descending chain of ideals*

$$\mathfrak{g} = \mathfrak{a}_0 \supseteqq \mathfrak{a}_1 \supseteqq \cdots \supseteqq \mathfrak{a}_m = \{0\}$$

such that $[\mathfrak{a}_i, \mathfrak{a}_i] \subseteqq \mathfrak{a}_{i+1}$ *for all* i $(0 \le i \le m-1)$.

PROOF. In order to show (1) $\Rightarrow$ (2), set $\mathfrak{a}_i = \mathfrak{g}^{(i)}$. Conversely, (2) $\Rightarrow$ (1) is obtained by proving $\mathfrak{a}_i \supseteqq \mathfrak{g}^{(i)}$ by induction on i. □

LEMMA 4.5. *Let* $\mathfrak{g}$ *be a Lie algebra and let* $\mathfrak{a}$, $\mathfrak{b}$ *be ideals of* $\mathfrak{g}$.

(1) *If* $\mathfrak{a}$ *and* $\mathfrak{g}/\mathfrak{a}$ *are solvable, then* $\mathfrak{g}$ *is solvable.*

(2) *If* $\mathfrak{a}$ *and* $\mathfrak{b}$ *are solvable, then* $\mathfrak{a} + \mathfrak{b} (= \{X + Y \mid X \in \mathfrak{a}, Y \in \mathfrak{b}\})$, *which is also an ideal of* $\mathfrak{g}$, *is solvable.*

PROOF. (1) Since $\mathfrak{g}/\mathfrak{a}$ is solvable, there exists an integer m such that $(\mathfrak{g}/\mathfrak{a})^{(m)} = \{0\}$. This implies $\mathfrak{g}^{(m)} \subseteqq \mathfrak{a}$. On the other hand, since $\mathfrak{a}$ is solvable, there is an integer m' such that $\mathfrak{a}^{(m')} = \{0\}$. Therefore $\mathfrak{g}^{(m+m')} = \{0\}$.

(2) The mapping of $\mathfrak{b}$ into $(\mathfrak{a}+\mathfrak{b})/\mathfrak{a}$ which assigns $Y+\mathfrak{a}$ to $Y \in \mathfrak{b}$ is linear, and moreover a Lie algebra homomorphism. It can be proved that this homomorphism induces a Lie algebra isomorphism of $\mathfrak{b}/(\mathfrak{a}\cap\mathfrak{b})$ onto $(\mathfrak{a}+\mathfrak{b})/\mathfrak{a}$. Therefore $\mathfrak{b}/(\mathfrak{a}\cap\mathfrak{b}) \cong (\mathfrak{a}+\mathfrak{b})/\mathfrak{a}$. Since $\mathfrak{b}$ is solvable, $\mathfrak{b}/(\mathfrak{a}\cap\mathfrak{b})$ is solvable by Example 4.14, and $\mathfrak{a}+\mathfrak{b}$ is also solvable by (1). □

DEFINITION 4.7. (1) By Lemma 4.5(2), there is a largest ideal among the solvable ideals of $\mathfrak{g}$, which is called the *radical* of $\mathfrak{g}$.

(2) If the radical of a Lie algebra $\mathfrak{g}$ is equal to $\{0\}$, then $\mathfrak{g}$ is called *semisimple.*

Let $\mathfrak{g}$ be a Lie algebra and let $\mathfrak{r}$ be its radical. As we can see from the definition, $\mathfrak{g}/\mathfrak{r}$ is semisimple.

EXAMPLE 4.15. Let $\mathfrak{g} = \mathfrak{gl}_2(k)$. The center $\mathfrak{z}$ of $\mathfrak{g}$ is the set $\{\alpha I \in \mathfrak{g} | \alpha \in k\}$ of scalar matrices and $\dim \mathfrak{z} = 1$. If we set $\mathfrak{s} = \{X \in \mathfrak{gl}_2(k) | \operatorname{tr}(X) = 0\}$, $\dim \mathfrak{s} = 3$ and $\mathfrak{g} = \mathfrak{z} \oplus \mathfrak{s}$ (as vector space). Moreover, $\mathfrak{s}$ is an ideal of $\mathfrak{g}$. If we regard $\mathfrak{s}$ as Lie algebra, we can show by direct computation that $[\mathfrak{s}, \mathfrak{s}] = \mathfrak{s}$, which implies that $\mathfrak{s}$ is not solvable. Therefore $\mathfrak{s}$ has no ideals except $\{0\}$ and $\mathfrak{s}$. (If there were such an ideal, $\mathfrak{s}$ would be solvable by the results above.) In particular, the radical of $\mathfrak{s}$ is $\{0\}$ and $\mathfrak{s}$ is semisimple. Since $\mathfrak{z}$ is solvable and $\mathfrak{g}/\mathfrak{z} \cong \mathfrak{s}$ is semisimple, $\mathfrak{z}$ is the radical of $\mathfrak{g}$).

(b) Linear Lie algebras and algebraic Lie algebras. Chevalley developed the theory of replicas in connection with the problem of characterizing the Lie algebras that correspond to algebraic groups. (See the remark below.) Then he showed that the theory of replicas could be applied to the general theory of Lie algebras (4), about which we shall give an introduction.

First, we define the notion of algebraic Lie algebras.

DEFINITION 4.8. (1) Let V be a k-vector space $\neq \{0\}$. A Lie subalgebra of the Lie algebra $\mathfrak{gl}(V)$ is called a *linear Lie algebra.*

(2) Let $\mathfrak{g}$ be a linear Lie algebra. If $\mathfrak{g}$ contains all replicas of A for all $A \in \mathfrak{g}$, then $\mathfrak{g}$ is called an *algebraic Lie algebra.*

(3) Let $\mathfrak{g}$ be a linear Lie algebra $(\mathfrak{g} \subseteqq \mathfrak{gl}(V))$. The intersection of all the algebraic Lie algebras $(\subseteq \mathfrak{gl}(V))$ that contain $\mathfrak{g}$ is called the *algebraic closure* of $\mathfrak{g}$ and is denoted by $\mathfrak{g}^{\mathrm{ac}}$.

EXAMPLE 4.16. Let $W_1 \supseteqq W_2$ be subspaces of the tensor space $T_q^p(V)$ of type (p, q). Define

$$\mathfrak{g} = \{X \in \mathfrak{gl}(V) | X_{p,q}(W_1) \subseteqq W_2\},$$

then $\mathfrak{g}$ is an algebraic Lie algebra (for the definition of $X_{p,q}$, see §2).

In fact, $\mathfrak{g}$ is a subspace of $\mathfrak{gl}(V)$. For $X \in \mathfrak{g}$, since $W_2 \subseteqq W_1$, we have $X_{p,q}(W_2) \subseteqq X_{p,q}(W_1) \subseteqq W_2$. Therefore, for $X, Y \in \mathfrak{g}$,

$$\begin{aligned}[X, Y]_{p,q}(W_1) &= X_{p,q}(Y_{p,q}(W_1)) - Y_{p,q}(X_{p,q}(W_1)) \\ &\subseteqq X_{p,q}(W_2) - Y_{p,q}(W_2) \subseteqq W_2,\end{aligned}$$

(4) C. Chevalley, *Algebraic Lie algebra*, Ann. of Math. **48** (1947), 91–100.

which shows $[X, Y] \in \mathfrak{g}$. Thus $\mathfrak{g}$ is a Lie subalgebra of $\mathfrak{gl}(V)$. By Corollary 2 to Theorem 4.1, $\mathfrak{g}$ is algebraic.

EXAMPLE 4.17. Let $(\mathfrak{A}, m)$ be an algebra. $\mathfrak{g} = \mathrm{Der}(\mathfrak{A})$ (Example 4.1, §1) is a Lie subalgebra of $\mathfrak{gl}(\mathfrak{A})$, therefore, it is a linear Lie algebra. Moreover, it is algebraic. In fact, by Example 4.9, §2, for $D \in \mathfrak{gl}(\mathfrak{A})$, we have that $D \in \mathfrak{g}$ if and only if $D_{1,2}(t_m) = 0$, where t_m is the tensor corresponding to the multiplication of $\mathfrak{A}$. Thus, by Example 4.16, where $W_1 = kt_m$ and $W_2 = \{0\}$, $\mathfrak{g}$ is algebraic.

REMARK. Let $k = \mathbf{C}$. A subgroup G of $\mathrm{GL}_n(\mathbf{C})$ is called algebraic if there exists a family $P_\lambda(x_{ij})$ $(\lambda \in \Lambda)$ of polynomials in n^2 variables x_{ij} which defines G. In other words, G is given by

$$G = \{g = (g_{ij}) \in \mathrm{GL}_n(\mathbf{C}) | P_\lambda(g_{ij}) = 0 \text{ for all } \lambda \in \Lambda\}.$$

From the theory of Lie groups, the Lie algebra of the complex Lie group $\mathrm{GL}_n(\mathbf{C})$ is $\mathfrak{gl}_n(\mathbf{C})$ and there is a correspondence between the (connected) complex Lie subgroups of $\mathrm{GL}_n(\mathbf{C})$ and the Lie subalgebras of $\mathfrak{gl}_n(\mathbf{C})$. By this correspondence, a Lie subalgebra corresponds to an algebraic subgroup if and only if it is an algebraic Lie subalgebra as defined above ([5]).

PROPOSITION 4.6. *Let $\mathfrak{g}$ be a linear Lie algebra.*

(1) *If $\mathfrak{a}$ is an ideal of $\mathfrak{g}$, then $\mathfrak{a}$ is an ideal of $\mathfrak{g}^{ac}$.*

(2) *We have $[\mathfrak{g}, \mathfrak{g}] = [\mathfrak{g}^{ac}, \mathfrak{g}^{ac}]$.*

PROOF. Identifying $T_{1,1}$ with $\mathfrak{gl}(V)$, we have

$$X_{1,1}(Y) = [X, Y] \qquad (X, Y \in \mathfrak{gl}(V))$$

(Lemma 4.2). For subspaces $\mathfrak{a}_1 \supseteqq \mathfrak{a}_2$ of $\mathfrak{gl}(V)$, define

$$\mathfrak{tr}(\mathfrak{a}_1, \mathfrak{a}_2) = \{X \in \mathfrak{gl}(V) | [X, \mathfrak{a}_1] \subseteqq \mathfrak{a}_2\},$$

then, by Example 4.16, $\mathfrak{tr}(\mathfrak{a}_1, \mathfrak{a}_2)$ is an algebraic Lie algebra.

(1) Since $\mathfrak{a}$ is an ideal of $\mathfrak{g}$, we have $[\mathfrak{g}, \mathfrak{a}] \subseteqq \mathfrak{a}$, namely, $\mathfrak{g} \subseteqq \mathfrak{tr}(\mathfrak{a}, \mathfrak{a})$. Since the right side is algebraic, $\mathfrak{g}^{ac} \subseteqq \mathfrak{tr}(\mathfrak{a}, \mathfrak{a})$, which shows that $[\mathfrak{g}^{ac}, \mathfrak{a}] \subseteqq \mathfrak{a}$. We use this argument in the following.

(2) Clearly, $[\mathfrak{g}, \mathfrak{g}] \subseteqq [\mathfrak{g}^{ac}, \mathfrak{g}^{ac}]$. We will show $[\mathfrak{g}^{ac}, \mathfrak{g}^{ac}] \subseteqq [\mathfrak{g}, \mathfrak{g}]$. Since $[\mathfrak{g}, \mathfrak{g}] = \mathfrak{g}^{(1)}$, we have that $\mathfrak{g} \subseteqq \mathfrak{tr}(\mathfrak{g}, \mathfrak{g}^{(1)})$ and $\mathfrak{g}^{ac} \subseteqq \mathfrak{tr}(\mathfrak{g}, \mathfrak{g}^{(1)})$, which implies $[\mathfrak{g}^{ac}, \mathfrak{g}] = [\mathfrak{g}, \mathfrak{g}^{ac}] \subseteqq \mathfrak{g}^{(1)}$. Therefore $\mathfrak{g} \subseteqq \mathfrak{tr}(\mathfrak{g}^{ac}, \mathfrak{g}^{(1)})$ and $\mathfrak{g}^{ac} \subseteqq \mathfrak{tr}(\mathfrak{g}^{ac}, \mathfrak{g}^{(1)})$. Namely $[\mathfrak{g}^{ac}, \mathfrak{g}^{ac}] \subseteqq \mathfrak{g}^{(1)}$. □

COROLLARY. *Let $\mathfrak{g}$ be a linear Lie algebra. Then $\mathfrak{g}$ is an ideal of $\mathfrak{g}^{ac}$ and $\mathfrak{g}^{ac}/\mathfrak{g}$ is a commutative Lie algebra. Moreover, we have $(\mathfrak{g}^{ac})^{(m)} = \mathfrak{g}^{(m)}$ and $(\mathfrak{g}^{ac})_m = \mathfrak{g}_m$. In particular, $\mathfrak{g}$ is commutative, nilpotent, or solvable if and only if $\mathfrak{g}^{ac}$ has the same property.*

([5]) For details, see the book, C. Chevalley, *Théorie des groupes de Lie*, tome 2, Hermann, Paris. But, in the book, the definition of replicas is different, and the result is not mentioned in this form.

PROOF. By (1), $\mathfrak{g}$ is an ideal of $\mathfrak{g}^{ac}$. Since $[\mathfrak{g}^{ac}, \mathfrak{g}^{ac}] = [\mathfrak{g}, \mathfrak{g}] \subseteqq \mathfrak{g}$, $[\mathfrak{g}^{ac}/\mathfrak{g}, \mathfrak{g}^{ac}/\mathfrak{g}] = \{0\}$. $(\mathfrak{g}^{ac})^{(1)} = \mathfrak{g}^{(1)}$ implies $(\mathfrak{g}^{ac})^{(m)} = \mathfrak{g}^{(m)}$ in general. Let us show $(\mathfrak{g}^{ac})_m = \mathfrak{g}_m$. For $m = 1$, $\mathfrak{g}_1 = \mathfrak{g}^{(1)}$. Assume that $(\mathfrak{g}^{ac})_{m-1} = \mathfrak{g}_{m-1}$. Then, since $\mathfrak{g} \subseteqq \mathfrak{tr}(\mathfrak{g}_{m-1}, \mathfrak{g}_m)$, $\mathfrak{g}^{ac} \subseteqq \mathfrak{tr}(\mathfrak{g}_{m-1}, \mathfrak{g}_m)$, which implies $[\mathfrak{g}^{ac}, \mathfrak{g}_{m-1}] \subseteqq \mathfrak{g}_m$. By assumption, the left side is equal to $(\mathfrak{g}^{ac})_m$ and we have $(\mathfrak{g}^{ac})_m = \mathfrak{g}_m$. □

(c) Engel's theorem and Lie's theorem. First, we prove Engel's Theorem (Proposition 4.7, Theorem 4.6) on Lie algebras consisting of nilpotent linear transformations. Then we discuss Lie's Theorem (Theorem 4.7) on solvable Lie algebras.

LEMMA 4.6. *Let $\mathfrak{g}$ be a linear Lie algebra contained in $\mathfrak{gl}(V)$. If $X \in \mathfrak{g}$ is a nilpotent linear transformation, $\mathrm{ad}(X)$ is a nilpotent element of $\mathfrak{gl}(\mathfrak{g})$.*

PROOF. If we identify $\mathfrak{gl}(V)$ with $T_1{}^1(V)$, then $\mathrm{ad}(X)$ is the restriction of $X_{1,1}$ to $\mathfrak{g}$. Therefore $\mathrm{ad}(X)$ is nilpotent by Lemma 4.4. □

PROPOSITION 4.7. *Let $\mathfrak{g}$ be a linear Lie algebra contained in $\mathfrak{gl}(V)$. If all elements of $\mathfrak{g}$ are nilpotent, then there is a nonzero element $v \in V$ such that $X(v) = 0$ for all $X \in \mathfrak{g}$.*

PROOF. We prove this assertion by induction on the dimension d of $\mathfrak{g}$. If $d = 0$, then $\mathfrak{g} = \{0\}$ and the result is obvious. Assume that $\dim \mathfrak{g} = d \geq 1$ and that the proposition is true for Lie algebras of dimensions $\leq (d-1)$. First, we shall show that, for a Lie subalgebra $\mathfrak{n}$ of $\mathfrak{g}$, where $\mathfrak{n} \subsetneqq \mathfrak{g}$, there exists a Lie subalgebra $\mathfrak{n}_1$ of $\mathfrak{g}$ with $\dim \mathfrak{n}_1 = \dim \mathfrak{n} + 1$ and $\mathfrak{n}$ an ideal of $\mathfrak{n}_1$. Now, the mapping which assigns $\mathrm{ad}(X) \in \mathfrak{gl}(\mathfrak{g})$ to $X \in \mathfrak{n}$ is a representation of $\mathfrak{n}$ on $\mathfrak{g}$ and, since $\mathrm{ad}(X)(\mathfrak{n}) \subseteqq \mathfrak{n}$, $\mathrm{ad}(X)$ induces a linear transformation $\psi(X)$ on the factor space $\mathfrak{g}/\mathfrak{n}$. Let $\psi(\mathfrak{n}) = \{\psi(X) | X \in \mathfrak{n}\}$. $\psi(\mathfrak{n})$ is a linear Lie algebra. Since $\mathrm{ad}(X)$ is nilpotent by Lemma 4.6, $\psi(X)$ is also nilpotent. Since $\dim \psi(\mathfrak{n}) \leq \dim \mathfrak{n} < d$, it follows by induction that there exists a $\overline{Y} \in \mathfrak{g}/\mathfrak{n}$ ($\overline{Y} \neq 0$) such that $\psi(\mathfrak{n})(\overline{Y}) = 0$, that is, a $Y \in \mathfrak{g}$ such that $Y \notin \mathfrak{n}$ and $[\mathfrak{n}, Y] \subseteqq \mathfrak{n}$. Set $\mathfrak{n}_1 = \mathfrak{n} + kY$, then $\mathfrak{n}_1$ satisfies the required condition.

If we repeat above procedure, starting from $\mathfrak{n}_0 = \{0\}$, we obtain a sequence:

$$\{0\} \subsetneqq \mathfrak{n}_1 \subsetneqq \mathfrak{n}_2 \subsetneqq \cdots \subsetneqq \mathfrak{g},$$

where $\dim \mathfrak{n}_i = \dim n_{i-1} + 1$, and $\mathfrak{n}_{i-1}$ is an ideal of $\mathfrak{n}_i$.

Let $\mathfrak{g} = \mathfrak{n}_d$, $\mathfrak{h} = \mathfrak{n}_{d-1}$ and define

$$W = \{v \in V | H(v) = 0 \text{ for all } H \in \mathfrak{h}\}.$$

By induction, we know that $W \neq \{0\}$. Moreover, for all $X \in \mathfrak{g}$, $X(W) \subseteqq W$. In fact, for $w \in W$ and $H \in \mathfrak{h}$,

$$H(X(w)) = [H, X](w) + X(H(w)) = 0.$$

Thus $X(w) \in W$.

Take $A \in \mathfrak{g} - \mathfrak{h}$. Then $\mathfrak{g} = \mathfrak{h} + kA$, A is nilpotent, and $A(W) \subseteqq W$. Therefore there is a $w \in W$ $(w \neq 0)$ such that $A(w) = 0$. This w is what we are looking for. □

THEOREM 4.6. *Let* $\mathfrak{g}$ *be a linear Lie algebra contained in* $\mathfrak{gl}(V)$. *Assume that all elements of* $\mathfrak{g}$ *are nilpotent linear transformations. Then,*

(1) $\mathfrak{g}$ *is a nilpotent Lie algebra.*

(2) *There exists a basis of* V *which satisfies the following condition: The matrix of an element of* $\mathfrak{g}$ *with respect to this basis is contained in* $\mathfrak{u}$ (= *the set of upper triangular matrices whose diagonal component are all* 0).

PROOF. We will prove (2). Then, since $\mathfrak{u}$ is a nilpotent Lie algebra (Example 4.13), (1) follows. Let v_1 be an element (whose existence is proved by Proposition 4.7) such that $X(v_1) = 0$ for all $X \in \mathfrak{g}$. Let W_1 be the 1-dimensional subspace generated by v_1. For all $X \in \mathfrak{g}$, $X(W_1) \subseteqq W_1$. Therefore X induces a linear transformation $\overline{X}$ of V/W_1. Since $\overline{X}$ is nilpotent it follows that for the Lie subalgebra $\{\overline{X} | X \in \mathfrak{g}\}$ of $\mathfrak{gl}(V/W_1)$ there exists $\overline{v}_2 \in V/W_1$ $(v_2 \in V)$ such that $\overline{X}(\overline{v}_2) = 0$ for all $\overline{X}$, i.e., $X(v_2) \in W_1$ for all $X \in \mathfrak{g}$. Let W_2 be the subspace generated by $\{v_1, v_2\}$, then we have $v_3 \in V - W_2$ such that $X(v_3) \in W_2$ for all $X \in \mathfrak{g}$. We can choose $v_4, \dots, v_n$ similarly. The basis $(v_1, \dots, v_n)$ has the required property. □

Without using the fact that the Lie algebra $\mathfrak{u}$ is nilpotent, we may prove (1) directly as follows. Consider the adjoint representation $X \mapsto \mathrm{ad}(X)$ of $\mathfrak{g}$. Since $\mathrm{ad}(X)$ is nilpotent, we can apply Proposition 4.7 to $\mathrm{ad}(\mathfrak{g}) \subseteqq \mathfrak{gl}(\mathfrak{g})$. Then, we know that there is an element $C \neq 0$ in $\mathfrak{g}$ such that $\mathrm{ad}(X)(C) = [X, C] = 0$ for all $X \in \mathfrak{g}$. Namely, the center $\mathfrak{z}_1$ of $\mathfrak{g}$ is $\mathfrak{z}_1 \neq \{0\}$. Next we will show that $\mathfrak{g}_1 = \mathfrak{g}/\mathfrak{z}_1$ is a linear Lie algebra and that all its elements are nilpotent. In fact, since the kernel of the adjoint representation is the center $\mathfrak{z}_1$, $\mathfrak{g}_1$ can be identified with $\mathrm{ad}(\mathfrak{g})$, which is linear and whose elements are all nilpotent. Similarly, we now that the center $\mathfrak{z}_2/\mathfrak{z}_1$ of $\mathfrak{g}_1$ is $\neq \{0\}$. By repeating this, it follows that $\mathfrak{g}$ is nilpotent by Proposition 4.4. By applying this result to $\mathfrak{u}$, we can show that $\mathfrak{u}$ is a nilpotent Lie algebra.

COROLLARY. *For a Lie algebra* $\mathfrak{g}$, $\mathfrak{g}$ *is nilpotent if and only if* $\mathrm{ad}(X)$ *is nilpotent for every* $X \in \mathfrak{g}$.

PROOF. If $\mathfrak{g}$ is nilpotent, then there is a sequence of ideals

$$\mathfrak{g} = \mathfrak{a}_0 \supseteqq \mathfrak{a}_1 \supseteqq \cdots \supseteqq \mathfrak{a}_m = \{0\}$$

with $[\mathfrak{g}, \mathfrak{a}_i] \subseteqq \mathfrak{a}_{i+1}$ (Proposition 4.4). Then we have, for $X \in \mathfrak{g}$, $\mathrm{ad}(X)(\mathfrak{g}) \subseteqq \mathfrak{a}_1$, $\mathrm{ad}(X)(\mathfrak{a}_1) \subseteqq \mathfrak{a}_2$, etc., which implies $(\mathrm{ad}(X))^m = 0$.

We will prove the converse. Let $\mathfrak{z}$ be the center of $\mathfrak{g}$. By the adjoint representation, $\mathfrak{g}/\mathfrak{z}$ is isomorphic to $\mathrm{ad}(\mathfrak{g}) \subseteqq \mathfrak{gl}(\mathfrak{g})$. By the theorem, $\mathrm{ad}(\mathfrak{g})$ is a nilpotent Lie algebra, thus there is an integer m such that $(\mathfrak{g}/\mathfrak{z})_m = \{0\}$, namely, $\mathfrak{g}_m \subseteqq \mathfrak{z}$. Then $\mathfrak{g}_{m+1} = \{0\}$. □

Next, we derive results leading to Lie's Theorem (Theorem 4.7).

LEMMA 4.7. *Let* $\mathfrak{g}$ *be a linear Lie algebra contained in* $\mathfrak{gl}(V)$, *and let* $\mathfrak{n}$ *be an ideal of* $\mathfrak{g}$ *such that all elements of* $\mathfrak{n}$ *are nilpotent. We regard the inclusion mapping* $\mathfrak{g} \to \mathfrak{gl}(V)$ *as representation of* $\mathfrak{g}$, *which we denote by* Φ. *Let* W *be a* $\mathfrak{g}$*-invariant subspace of* V. *If the representation* Φ_W *is completely reducible,* $\Phi_W(\mathfrak{n}) = \{(N|W)|N \in \mathfrak{n}\}$ *is equal to* $\{0\}$, *where* $N|W$ *denotes the restriction of* N *to* W.

PROOF. Let $W' = \{v \in W | N(v) = 0 \text{ for all } N \in \mathfrak{n}\}$. It suffices to show that $W' = W$. For all $X \in \mathfrak{g}$, $N \in \mathfrak{n}$,

$$N(X(v)) = X(N(v)) + [N, X](v).$$

Therefore W' is a $\mathfrak{g}$-invariant subspace. Since Φ is completely reducible, there exists a $\mathfrak{g}$-invariant subspace W'' such that $W = W' \oplus W''$ (direct sum). Suppose that $W'' \neq \{0\}$. Considering $\Phi_{W''}$ as a representation of $\mathfrak{n}$, we can conclude that, by Proposition 4.7, there is an element $w \in W''$ ($w \neq 0$) such that $Nw = 0$ for all $N \in \mathfrak{n}$. This implies $w \in W'$, a contradiction. □

PROPOSITION 4.8. *Let* $\mathfrak{g}$ *be a linear Lie algebra contained in* $\mathfrak{gl}(V)$. *Assume that the representation of* $\mathfrak{g}$ *defined by the inclusion mapping* $\mathfrak{g} \to \mathfrak{gl}(V)$ *is completely reducible. Then the radical of* $\mathfrak{g}$ *coincides with the center of* $\mathfrak{g}$.

PROOF. We denote the center and the radical of $\mathfrak{g}$ by $\mathfrak{z}$ and $\mathfrak{r}$ respectively. First we prove that $\mathfrak{z} \cap [\mathfrak{g}, \mathfrak{g}] = \{0\}$. $\mathfrak{z} \cap [\mathfrak{g}, \mathfrak{g}]$ is an ideal of $\mathfrak{g}$, and each of its elements is nilpotent. In fact, let $Z = \sum [X_i, Y_i] \in \mathfrak{z} \cap [\mathfrak{g}, \mathfrak{g}]$ $(X_i, Y_i \in \mathfrak{g})$ and let Z' be a replica of Z. Since Z' is a polynomial in Z, $[Z, X_i] = [Z, Y_i] = 0$ (recall that $Z \in \mathfrak{z}$) implies that $[Z', X_i] = [Z', Y_i] = 0$. then

$$\operatorname{tr}(Z', Z) = \sum_i \operatorname{tr}(Z'[X_i, Y_i]) = \sum_i \operatorname{tr}([Z', X_i]Y_i) = 0.$$

By Theorem 4.5, Z is nilpotent. Using Lemma 4.7 for $W = V$, we obtain $\mathfrak{z} \cap [\mathfrak{g}, \mathfrak{g}] = \{0\}$.

Suppose that $\mathfrak{r} \supsetneqq \mathfrak{z}$. Since r is solvable, there is an integer m such that $(\mathfrak{r}/\mathfrak{z})^{(m-1)} \neq \{0\}$ and $(\mathfrak{r}/\mathfrak{z})^{(m)} = \{0\}$. Set $(\mathfrak{r}/\mathfrak{z})^{(m-1)} = \mathfrak{a}/\mathfrak{z}$, then $[\mathfrak{a}, \mathfrak{a}] \subseteq \mathfrak{z} \cap [\mathfrak{g}, \mathfrak{g}] = \{0\}$. Now, we shall show that every element of $[\mathfrak{a}, \mathfrak{g}]$ is nilpotent. Let $W = \sum_i [A_i, X_i] \in [\mathfrak{a}, \mathfrak{g}]$ $(A_i \in \mathfrak{a},\ X_i \in \mathfrak{g})$ and W' be a replica of W. Since $W \in \mathfrak{a}$, $[W, A_i] = 0$ and $[W', A_i] = 0$. Then

$$\operatorname{tr}(W'W) = \sum_i \operatorname{tr}(W'[A_i, X_i]) = \sum_i \operatorname{tr}([W', A_i]X_i) = 0.$$

Therefore W is nilpotent and $[\mathfrak{a}, \mathfrak{g}] = \{0\}$ as above. This implies $\mathfrak{a} \subseteqq \mathfrak{z}$ and contradicts $\mathfrak{a}/\mathfrak{z} \neq \{0\}$. □

EXAMPLE 4.18. Let $\mathfrak{g} = \mathfrak{gl}(V)$. Since the representation $X \mapsto X \in \mathfrak{gl}(V)$ of $\mathfrak{g}$ is irreducible, it is completely reducible. Therefore, the radical of $\mathfrak{g}$ coincides with the center of $\mathfrak{g}$, which is clearly the set of scalar transformations (See Example 4.15).

THEOREM 4.7. *Let* $\mathfrak{g}$ *be a solvable Lie algebra and let* Φ *be an irreducible representation of* $\mathfrak{g}$. *Then the image* $\Phi(\mathfrak{g})$ *is commutative.*

PROOF. Since Φ is irreducible, it is completely reducible and $\Phi(\mathfrak{g})$ satisfies the condition of Proposition 4.8. Since $\mathfrak{g}$ is solvable, $\Phi(\mathfrak{g})$ is also solvable. Therefore the radical of $\Phi(\mathfrak{g})$ is $\Phi(\mathfrak{g})$ itself, which is the center of $\Phi(\mathfrak{g})$. In other words, $\Phi(\mathfrak{g})$ is commutative. □

COROLLARY 1. *If* $k = \mathbf{C}$, *then the representation space of an irreducible representation of a solvable Lie algebra is of dimension* 1.

PROOF. Let Φ be an irreducible representation of $\mathfrak{g}$ with representation space V. By the theorem, the elements of $\Phi(\mathfrak{g})$ commute each other. Let α be an eigenvalue of $A \in \Phi(\mathfrak{g})$. Then, for all $X \in \Phi(\mathfrak{g})$,

$$X(A - \alpha I) = (A - \alpha I)X$$

(I is the identity transformation of V). Thus $A'(V)$, where $A' = A - \alpha I$, is a $\mathfrak{g}$-invariant subspace. From the irreducibility of Φ, $A'(V) = \{0\}$ or $A'(V) = V$. Since α is an eigenvalue, it follows that $A'(V) = \{0\}$, which implies $A' = A - \alpha I = 0$, $A = \alpha I$. Since Φ is irreducible, we have that $\dim V = 1$. □

COROLLARY 2. *Let* $k = \mathbf{C}$, *let* $\mathfrak{g}$ *be a solvable Lie algebra, and let* Φ *be a representation of* $\mathfrak{g}$ *with representation space* V $(\neq \{0\})$.

(1) *There is an element* $w \in W$ $(w \neq 0)$ *that is an eigenvector for all* $\Phi(X)$, $X \in \mathfrak{g}$.

(2) *There is a basis for* V *which satisfies the following condition: With respect to this basis, the matrix* $\Phi(X)$ *is an upper triangle matrix for all* $X \in \mathfrak{g}$.

PROOF. Let W be a $\mathfrak{g}$-invariant subspace of V $(W \neq \{0\})$, whose dimension is minimal. Since Φ_W is irreducible, $\dim W = 1$, and any nonzero $w \in W$ satisfies (1).

(2) We use the same argument as we used to deduce Theorem 4.6(2) from Proposition 4.7. Instead of Proposition 4.7, we use (1) and set $v_1 = w$. If $\dim V > 1$, we apply (1) again to the representation Φ_{V/W_1} on V/W_1, where W_1 is the subspace generated by v_1, and we know that there is a $\overline{v}_2 \neq 0$ in V/W_1 which is an eigenvector for all $\Phi_{V/W}(X)$ $(X \in \mathfrak{g})$. Let W_2 be the subspace generated by $\{v_1, v_2\}$ and choose an element $\overline{v}_3 \neq 0$ in V/W_2 $(v_3 \in V)$ that is an eigenvector of all $\Phi_{V/W_2}(X)$. Repeating this procedure, we construct a basis $(v_1, \dots, v_n)$ that satisfies the requirement. □

REMARK. In Corollaries 1 and 2, if k is an algebraically closed field of characteristic 0, then the same results hold.

(d) Cartan's criterion. Here, we prove Cartan's criterion for the semisimplicity of a Lie algebra. Let $\mathfrak{g}$ be a linear Lie algebra contained in $\mathfrak{gl}(V)$. Define

$$\mathfrak{n}(\mathfrak{g}) = \{X \mid X \in \mathfrak{g},\ \operatorname{tr}(XY) = 0 \text{ for all } Y \in \mathfrak{g}\}.$$

Then $\mathfrak{n}(\mathfrak{g})$ is an ideal of $\mathfrak{g}$. In fact, for $X \in \mathfrak{n}(\mathfrak{g})$ and $Y, Z \in \mathfrak{g}$,

$$\operatorname{tr}([X, Y]Z) = \operatorname{tr}(X[Y, Z]) = 0,$$

which implies $[X, Y] \in \mathfrak{n}(\mathfrak{g})$. We have the next proposition.

PROPOSITION 4.9. *If $\mathfrak{g}$ is algebraic, every element of $\mathfrak{n} = \mathfrak{n}(\mathfrak{g})$ is nilpotent. In general, if $\mathfrak{g}$ is a linear Lie algebra, then every element of $[\mathfrak{n}, \mathfrak{g}]$ is nilpotent.*

PROOF. Let $\mathfrak{g}$ be an algebraic Lie algebra. Let $X \in \mathfrak{n}$ and $X \twoheadrightarrow Y$. Then $Y \in \mathfrak{g}$ and $\operatorname{tr}(XY) = 0$. Therefore X is nilpotent. In general, let $\mathfrak{g}$ be a linear Lie algebra, let $\mathfrak{g}^{ac}$ be its algebraic closure, and let $\mathfrak{n}(\mathfrak{g}^{ac}) = \mathfrak{n}_1$. It suffices to show $[\mathfrak{n}, \mathfrak{g}] \subseteqq \mathfrak{n}_1$. Let $X \in \mathfrak{g}^{ac}$, $X_i \in \mathfrak{g}$, and $N_i \in \mathfrak{n}$. Then

$$\operatorname{tr}\left(\left(\sum_i [N_i, X_i]\right) X\right) = \operatorname{tr}\left(\sum_i N_i[X_i, X]\right),$$

where $[X_i, X] \in [\mathfrak{g}, \mathfrak{g}^{ac}] \subseteqq [\mathfrak{g}^{ac}, \mathfrak{g}^{ac}] = [\mathfrak{g}, \mathfrak{g}] \subseteqq \mathfrak{g}$. Since $N_i \in \mathfrak{n}$, the right side equals zero. Therefore $\sum [N_i, X_i] \in \mathfrak{n}_1$, as required. □

COROLLARY 1. *Let $\mathfrak{g}$ be a linear Lie algebra. If $\mathfrak{g} = \mathfrak{n}(\mathfrak{g})$, then $\mathfrak{g}$ is solvable.*

PROOF. Since $[\mathfrak{g}, \mathfrak{g}]$ is a nilpotent Lie algebra, $\mathfrak{g}$ is solvable. □

COROLLARY 2. *Let $\mathfrak{g}$ be a semisimple Lie algebra and Φ a one-to-one representation of $\mathfrak{g}$. Define a bilinear form f on $\mathfrak{g} \times \mathfrak{g}$ by $f(X, Y) = \operatorname{tr}(\Phi(X)\Phi(Y))$. Then f is nondegenerate.*

PROOF. Choose a basis $(X_1, \dots, X_d)$ of $\mathfrak{g}$. It suffices to prove $\Delta_f = \det(f(X_i, X_j)) \neq 0$. Suppose that $\Delta_f = 0$. Then there exists an element $X \neq 0$ of $\mathfrak{g}$ such that $f(X_i, X) = 0$ for all i. Denote the images of $\mathfrak{g}$ and $\mathfrak{n}(\mathfrak{g})$ under Φ by $\bar{\mathfrak{g}}$ and $\bar{\mathfrak{n}}$ respectively. Since Φ is one-to-one, $\Phi(X) \neq 0$. And $\Phi(X) \in \bar{\mathfrak{n}}$, because $f(X_i, X) = \operatorname{tr}(\Phi(X_i)\Phi(X)) = 0$. Since $\mathfrak{n}(\bar{\mathfrak{n}}) = \bar{\mathfrak{n}}$, $\bar{\mathfrak{n}}$ is a solvable and an ideal of $\bar{\mathfrak{g}}$. Therefore the radical of $\bar{\mathfrak{g}} \neq \{0\}$. This contradicts the fact that $\mathfrak{g}$ $(\cong \bar{\mathfrak{g}})$ is semisimple. □

DEFINITION 4.9. Let $\mathfrak{g}$ be a Lie algebra. The bilinear form B on $\mathfrak{g} \times \mathfrak{g}$ defined by

$$B(X, Y) = \operatorname{tr}(\operatorname{ad}(X)\operatorname{ad}(Y)) \qquad (X, Y \in \mathfrak{g})$$

is called the *Killing form* of $\mathfrak{g}$.

EXERCISE 1. Let $\mathfrak{g}$ be a Lie algebra and let $\mathfrak{a}$ be an ideal of $\mathfrak{g}$. Denote by B and $B_{\mathfrak{a}}$ the Killing form of $\mathfrak{g}$ and $\mathfrak{a}$ respectively. Then, prove $B_{\mathfrak{a}}(X, Y) = B(X, Y)$ for $X, Y \in \mathfrak{a}$.

EXAMPLE 4.19. Let $\mathfrak{g} = \mathfrak{gl}_n(k)$. We have, for $X, Y \in \mathfrak{g}$,

$$B(X, Y) = 2n\operatorname{tr}(XY) - 2\operatorname{tr}(X)\operatorname{tr}(Y).$$

Define a subspace $\mathfrak{g}_1$ of $\mathfrak{g}$ by

$$\mathfrak{g}_1 = \{X \in \mathfrak{gl}_n(k) \mid \operatorname{tr}(X) = 0\}.$$

Then $\mathfrak{g}_1$ is an ideal of $\mathfrak{g}$. The Lie algebra $\mathfrak{g}_1$ is denoted by $\mathfrak{sl}_n(k)$. For $X, Y \in \mathfrak{sl}_n(k)$, $B(X, Y) = 2n\,\mathrm{tr}(XY)$.

EXERCISE 2. Verify the above facts.

THEOREM 4.8 (Cartan). *Let $\mathfrak{g}$ be a Lie algebra. $\mathfrak{g}$ is semisimple if and only if its Killing form is nondegenerate.*

PROOF. If $\mathfrak{g}$ is semisimple, then the center of $\mathfrak{g} = \{0\}$. Therefore the adjoint representation is one-to-one. By Corollary 2 to Proposition 4.9, the Killing form is nondegenerate.

Conversely, suppose that $\mathfrak{g}$ is not semisimple, namely, the radical $\mathfrak{r}$ of $\mathfrak{g} \neq \{0\}$. Since $\mathfrak{r}$ is solvable, there exists an integer m such that $\mathfrak{r}^{(m-1)} \neq \{0\}$ and $\mathfrak{r}^{(m)} = \{0\}$. Set $\mathfrak{a} = \mathfrak{r}^{(m-1)}$. $\mathfrak{a}$ is an ideal of $\mathfrak{g}$ and $[\mathfrak{a}, \mathfrak{a}] = \{0\}$. For all $A \in \mathfrak{a}$, $X \in \mathfrak{g}$, we have

$$\mathrm{ad}(A)\,\mathrm{ad}(X)(\mathfrak{g}) = [A, [X, \mathfrak{g}]] \subseteqq [A, \mathfrak{g}] \subseteqq \mathfrak{a}.$$

Therefore

$$(\mathrm{ad}(A)\,\mathrm{ad}(X))^2(\mathfrak{g}) \subseteqq \mathrm{ad}(A)\,\mathrm{ad}(\mathfrak{g})(\mathfrak{a}) = \{0\}.$$

That is, $\mathrm{ad}(A)\,\mathrm{ad}(X)$ is nilpotent and $\mathrm{tr}(\mathrm{ad}(A)\,\mathrm{ad}(X)) = B(A, X) = 0$. Since $\mathfrak{a} \neq \{0\}$, there is an element $A \neq 0$ such that $B(A, X) = 0$ for all $X \in \mathfrak{g}$ and B is degenerate. □

EXAMPLE 4.20. By Example 4.19, we can prove that the Killing form of $\mathfrak{sl}_n(k)$ $(n \geq 2)$ is nondegenerate. Therefore $\mathfrak{sl}_n(k)$ $(n \geq 2)$ is semisimple.

By Cartan's Theorem, we can easily prove the following properties of a semisimple Lie algebra $\mathfrak{g}$.

(1) Let $\mathfrak{a}$ be an ideal of $\mathfrak{g}$. Define $\mathfrak{a}^{\perp} = \{Y \in \mathfrak{g} | B(X, Y) = 0$ for all $X \in \mathfrak{g}\}$. Then $\mathfrak{a}^{\perp}$ is an ideal, $\mathfrak{a} \cap \mathfrak{a}^{\perp} = \{0\}$, and $\mathfrak{g} = \mathfrak{a} \oplus \mathfrak{a}^{\perp}$ (direct sum). In other words, the adjoint representation of a semisimple Lie algebra is completely reducible. Therefore $\mathfrak{g}$ $(\neq \{0\})$ is decomposed into a direct sum of irreducible $\mathfrak{g}$-invariant subspaces (i.e., minimal ideals of $\mathfrak{g}$, $\neq \{0\}$) $\mathfrak{a}_1, \dots, \mathfrak{a}_r$:

$$\mathfrak{g} = \mathfrak{a}_1 \oplus \cdots \oplus \mathfrak{a}_r.$$

The $\mathfrak{a}_i$ are not commutative. $\mathfrak{a}_i$ has no ideals except $\{0\}$ and $\mathfrak{a}_i$ itself. The decomposition into a direct sum of minimal ideals is unique up to order. That is, if $\mathfrak{g} = \mathfrak{b}_1 \oplus \cdots \oplus \mathfrak{b}_s$ (direct sum) where $\mathfrak{b}_i$ is a minimal ideal, then we have $r = s$ and $\mathfrak{b}_i = \mathfrak{a}_i$ for all i, after changing the indices properly. Moreover $\mathfrak{g} = [\mathfrak{g}, \mathfrak{g}]$.

(2) Let D be a derivation of $\mathfrak{g}$. There is an $A \in \mathfrak{g}$ such that $D = \mathrm{ad}(A)$.

EXERCISE 3. Prove the above assertions (1) and (2).

There are other general results on Lie algebras. For example, Levi's Theorem (if $\mathfrak{g}$ is a Lie algebra with radical $\mathfrak{r}$, there exists a semisimple subalgebra $\mathfrak{s}$ of $\mathfrak{g}$ such that $\mathfrak{g} = \mathfrak{r} \oplus \mathfrak{s}$ (direct sum)) and Weyl's Theorem on complete reducibility (of representations of semisimple Lie algebras). These two results can also be presented as applications of the theory of replicas ([6]). But, since they require much more preparation and we have already presented significant results about Lie algebras, we stop at this point.

Exercises

1. Define a Lie subalgebra $\mathfrak{g}$ of $\mathfrak{gl}_2(k)$ by :

$$\mathfrak{g} = \left\{ \begin{bmatrix} a & b \\ 0 & c \end{bmatrix} \middle| \, a , b , c \in k \right\} .$$

(1) Choose a basis for $\mathfrak{g}$. Let A be the matrix of a linear transformation F of $\mathfrak{g}$. Give necessary and sufficient conditions with respect to A for F to be a derivation of $\mathfrak{g}$.

(1) Using (1), give the dimension of the Lie algebra $\mathrm{Der}(\mathfrak{g})$. Compute the dimension of the ideal $\mathrm{ad}(\mathfrak{g})$ of $\mathrm{Der}(\mathfrak{g})$.

2. Let $V = \mathbf{C}^3$ (the complex vector space consisting of all the column vectors in $\mathbf{C}$ of dimension 3) and let F_i be the linear transformation $(V \ni v \mapsto A_i v \in V)$ of V defined by the matrix A_i $(i = 1, 2, 3, 4)$. Give a basis for the vector space of replicas of F_i.

$$(1)\ A_1 = \begin{bmatrix} 0 & 1 & 0 \\ 0 & 0 & 1 \\ 1 & 0 & 0 \end{bmatrix}, \qquad (2)\ A_2 = \begin{bmatrix} 1 & 0 & 0 \\ 0 & \sqrt{2} & 0 \\ 0 & 0 & 1+\sqrt{2} \end{bmatrix},$$

$$(3)\ A_3 = \begin{bmatrix} \sqrt{2} & 1 & 0 \\ 0 & \sqrt{2} & 0 \\ 0 & 0 & 1 \end{bmatrix}, \qquad (4)\ A_4 = \begin{bmatrix} 6 & -3 & -2 \\ 4 & -1 & -2 \\ 3 & -2 & 0 \end{bmatrix}.$$

3. Let $\mathfrak{g}$ be defined as follows:

$$\mathfrak{g} = \left\{ \left[\begin{array}{c|c} A_1 & B \\ \hline 0 & A_2 \end{array} \right] \in \mathfrak{gl}_n(k) \;\middle|\; \begin{array}{l} A_1 \in \mathfrak{gl}_{n_1}(k),\ A_2 \in \mathfrak{gl}_{n_2}(k), \\ B \in M(n_1, n_2; k) \\ (n = n_1 + n_2). \end{array} \right\}$$

(1) Prove that $\mathfrak{g}$ is a Lie subalgebra of $\mathfrak{gl}_n(k)$.
(2) Prove that $\mathfrak{g}$ is an algebraic Lie algebra.
(3) In general, show that the same assertions for

$$\left\{ \begin{bmatrix} A_1 & & & * \\ & A_2 & & \\ & & \ddots & \\ 0 & & & A_r \end{bmatrix} \in \mathfrak{gl}_n(k) \;\middle|\; \begin{array}{l} A_i \in \mathfrak{gl}_{n_i}(k),\ i = 1, \dots, r \\ (n = n_1 + \cdots + n_r). \end{array} \right\}$$

4. (1) Let $\mathfrak{a}$ be an ideal of a linear Lie algebra $\mathfrak{g}$. Prove that $[\mathfrak{g}^{\mathrm{ac}}, \mathfrak{a}^{\mathrm{ac}}] \subseteqq \mathfrak{a}$.

([6]) These are explained in the following articles: Nagayoshi Iwahori and Ichiro Satake, *On replicas of matrices* and *On structures of Lie algebras* (I) (II) (III) (Japanese) Mathematical Report, College of Arts and Sciences, University of Toyko (1950–52). The exposition in this book also depends on these articles.

(2) Let $\mathfrak{g}$ be an algebraic Lie algebra. Prove that the radical of $\mathfrak{g}$ is also algebraic.

5. Let $\mathfrak{g}$ be defined as follows:

$$\mathfrak{g} = \left\{ \begin{bmatrix} a_{11} & a_{12} & a_{13} \\ a_{21} & a_{22} & a_{23} \\ 0 & 0 & 0 \end{bmatrix} \middle| a_{ij} \in k \right\}.$$

(1) Prove that $\mathfrak{g}$ is a Lie subalgebra of $\mathfrak{gl}_3(k)$ and that it is algebraic.
(2) Compute the Killing form of $\mathfrak{g}$.
(3) Determine the radical of $\mathfrak{g}$.

6. For $\mathfrak{g}$ given in exercise 3, determine the ideal

$$\mathfrak{n}(\mathfrak{g}) = \{X | X \in \mathfrak{g},\ \operatorname{tr}(XY) = 0 \text{ for all } Y \in \mathfrak{g}\}.$$

7. Let $\mathfrak{g}$ be a Lie algebra and A an associative algebra with the unit element. We can regard A as a Lie algebra by Example 4.2, §1. Then, a Lie algebra homomorphism of $\mathfrak{g}$ into A is called an α-mapping. Namely, a linear mapping $\sigma: \mathfrak{g} \to A$ is an α-mapping if

$$\sigma([X, Y]) = \sigma(X)\sigma(Y) - \sigma(Y)\sigma(X)$$

for all $X, Y \in \mathfrak{g}$.

A pair $(\mathfrak{U}, \sigma_0)$ consisting of an associative algebra $\mathfrak{U}$ with unit element and an α-mapping σ_0 of $\mathfrak{g}$ into $\mathfrak{U}$ is called a *universal enveloping algebra* of $\mathfrak{g}$ if is satisfies the following condition (U):

(U) For any pair (A, F) consisting of an associative algebra A with unit element and an α-mapping F of $\mathfrak{g}$ into A, there is a unique associative algebra homomorphism $\widetilde{F}$ of $\mathfrak{U}$ into A such that $\widetilde{F} \circ \sigma_0 = F$.

For a Lie algebra $\mathfrak{g}$, prove the existence and the uniqueness of a universal enveloping algebra.

8. Let V be a vector space over $\mathbf{C}$.

(1) Prove that any linear transformation $A \in \mathfrak{gl}(V)$ can be expressed in the form:

$$A = A_{(0)} + \lambda_1 A_{(1)} + \cdots + \lambda_k A_{(k)},$$

where $A_{(0)}$ is the nilpotent part of A, $A_{(i)}$ $(i = 1, \ldots, k)$ is a semisimple linear transformation whose eigenvalues are all rational numbers, $\{\lambda_1, \ldots, \lambda_k\}$ is linearly independent over $\mathbf{Q}$, and the $A_{(j)}$ $(j = 0, 1, \ldots, k)$ commute with each other.

(2) Prove that the vector space of replicas of A is generated by $A_{(0)}$, $A_{(1)}, \ldots, A_{(k)}$.

References for the English Edition

1. W. H. Greub, *Multilinear algebra*, Grundlehren Math. Wiss., Bd. 136, Springer-Verlag, Berlin and New York, 1967.

In this book, tensor algebras are discussed at great length. This is the second volume of Greub's text books on linear algebra; the first one is, *Linear Algebra*, 3rd ed., Grundlehren Math. Wiss., Bd. 97, Springer-Verlag, Berlin and New York, 1967.

In Bourbaki's series of Éléments de Mathématique, tensor algebra is treated in:

2. N. Bourbaki, *Algèbre*, Hermann, Paris, 1970, Chapters 2 and 3.

Though there are no books in Japanese which are written about the same topics as the present volume, there are several books on linear algebra or algebra, parts of which are devoted to tensor algebras, e.g.:

3. I. Satake, *Linear algebra*, Marcel Dekker, New York, 1975.

The Japanese edition was published by Shokabo in 1973. A large part of the content of the present volume is explained neatly in Chapter V. A description of group representations is also included.

4. N. Iwahori, *Vector calculus*, Shokabo, Tokyo, 1960. (Japanese)

Tensors of type $(p\,,\,q)$ $(p+q \leq 2)$ on $\mathbf{R}^3$ are explained in detail. There are many examples and exercises.

In some books on differential geometry, tensor algebra and its relation to geometry are explained as preliminaries.

5. S. Sternberg, *Lectures on differential geometry*, Prentice Hall, Englewood Cliffs, NJ, 1964 (republished by Chelsea in 1983).

6. S. Kobayashi and K. Nomizu, *Foundations of differential geometry*, vol. 1, John Wiley & Sons, New York, 1963.

We consulted the above books during the preparation of the present volume.

The following books appeared more recently.

7. M. Marcus, *Finite dimensional multilinear algebra*, Parts I, II, Marcel Dekker, New York, 1973 (Part I) and 1975 (Part II).

8. L. Schwartz, *Les tenseurs*, Actualités scientifiques et industrielles, 1376, Hermann, Paris, 1975.

9. S. Lang, *Algebra*, 2nd ed., Addison-Wesley, Reading, MA, 1984.

Section 4.3 contains only the beginning of the theory of Lie algebras. For details, see e.g.

10. Y. Matsushima, *Theory of Lie algebras*, Kyoritsu Shuppan, Tokyo, 1956. (Japanese)

11. N. Iwahori, *Theory of Lie groups* II, Iwanami Shoten, Tokyo, 1957.

12. N. Jacobson, *Lie Algebras*, Interscience, New York, 1962 (republished by Dover).

13. J. E. Humphreys, *Introduction to Lie algebras and representation theory*, 2nd printing, revised. Graduated Texts in Math., vol. 9, Springer-Verlag, New York, 1972.

Subject Index

Recent Titles in This Series

(Continued from the front of this publication)

73 **Yu. G. Reshetnyak,** Space mappings with bounded distortion, 1989
72 **A. V. Pogorelev,** Bendings of surfaces and stability of shells, 1988
71 **A. S. Markus,** Introduction to the spectral theory of polynomial operator pencils, 1988
70 **N. I. Akhiezer,** Lectures on integral transforms, 1988
69 **V. N. Saliĭ,** Lattices with unique complements, 1988
68 **A. G. Postnikov,** Introduction to analytic number theory, 1988
67 **A. G. Dragalin,** Mathematical intuitionism: Introduction to proof theory, 1988
66 **Ye Yan-Qian,** Theory of limit cycles, 1986
65 **V. M. Zolotarev,** One-dimensional stable distributions, 1986
64 **M. M. Lavrent′ev, V. G. Romanov, and S. P. Shishat·skiĭ,** Ill-posed problems of mathematical physics and analysis, 1986
63 **Yu. M. Berezanskiĭ,** Selfadjoint operators in spaces of functions of infinitely many variables, 1986
62 **S. L. Krushkal′, B. N. Apanasov, and N. A. Gusevskiĭ,** Kleinian groups and uniformization in examples and problems, 1986
61 **B. V. Shabat,** Distribution of values of holomorphic mappings, 1985
60 **B. A. Kushner,** Lectures on constructive mathematical analysis, 1984
59 **G. P. Egorychev,** Integral representation and the computation of combinatorial sums, 1984
58 **L. A. Aĭzenberg and A. P. Yuzhakov,** Integral representations and residues in multidimensional complex analysis, 1983
57 **V. N. Monakhov,** Boundary-value problems with free boundaries for elliptic systems of equations, 1983
56 **L. A. Aĭzenberg and Sh. A. Dautov,** Differential forms orthogonal to holomorphic functions or forms, and their properties, 1983
55 **B. L. Roždestvenskiĭ and N. N. Janenko,** Systems of quasilinear equations and their applications to gas dynamics, 1983
54 **S. G. Kreĭn, Ju. I. Petunin, and E. M. Semenov,** Interpolation of linear operators, 1982
53 **N. N. Čencov,** Statistical decision rules and optimal inference, 1981
52 **G. I. Èskin,** Boundary value problems for elliptic pseudodifferential equations, 1981
51 **M. M. Smirnov,** Equations of mixed type, 1978
50 **M. G. Kreĭn and A. A. Nudel′man,** The Markov moment problem and extremal problems, 1977
49 **I. M. Milin,** Univalent functions and orthonormal systems, 1977
48 **Ju. V. Linnik and I. V. Ostrovskiĭ,** Decomposition of random variables and vectors, 1977
47 **M. B. Nevel′son and R. Z. Has′minskiĭ,** Stochastic approximation and recursive estimation, 1976
46 **N. S. Kurpel′,** Projection-iterative methods for solution of operator equations, 1976
45 **D. A. Suprunenko,** Matrix groups, 1976
44 **L. I. Ronkin,** Introduction to the theory of entire functions of several variables, 1974
43 **Ju. L. Daleckiĭ and M. G. Kreĭn,** Stability of solutions of differential equations in Banach space, 1974
42 **L. D. Kudrjavcev,** Direct and inverse imbedding theorems, 1974
41 **I. C. Gohberg and I. A. Fel′dman,** Convolution equations and projection methods for their solution, 1974

(See the AMS catalog for earlier titles)